Computer Graphics:
Systems and Applications

Managing Editor: J. L. Encarnação

Editors: K. Bø   J. D. Foley   R. A. Guedj
P. J. W. ten Hagen   F. R. A. Hopgood   M. Hosaka
M. Lucas   A. G. Requicha

Springer
Berlin
Heidelberg
New York
Barcelona
Budapest
Hong Kong
London
Milan
Paris
Santa Clara
Singapore
Tokyo

Fan Dai (Ed.)

Virtual Reality
for Industrial Applications

With 60 Figures,
16 in Colour

 Springer

Editor

Dr. Fan Dai
ABB Corporate Research
Speyerer Straße 4
D-69115 Heidelberg
Germany

ISBN-13: 978-3-642-46849-0   e-ISBN-13: 978-3-642-46847-6
DOI: 10.1007/978-3-642-46847-6

Library of Congress Cataloging-in-Publication Data

Virtual reality for industrial applications / Fan Dai, ed.
p. cm. – (Computer graphics – systems and applications)
Includes bibliographical references and index.
ISBN-13: 978-3-642-46849-0
1. Human-computer interaction.   2. Virtual reality.   3. Computer-aided engineering.
I. Dai, Fan, 1967– .   II. Series.
QA76.9.H85V582   1998   670'.285'6–dc21   97-43400   CIP

© Springer-Verlag Berlin Heidelberg 1998
Softcover reprint of the hardcover 1st edition 1998

The use of general descriptive names, registered names, trademarks, etc. in this publication does not imply, even in the absence of a specific statement, that such names are exempt from the relevant protective laws and regulations and therefore free for general use.

Cover Design: Künkel + Lopka Werbeagentur, Heidelberg
Typesetting: Camera ready by editor

SPIN: 10637574   33/3142 - 5 4 3 2 1 0 - Printed on acid-free paper

Preface

Just a few years ago, virtual reality was regarded as more a toy than a tool. Today, however, it is becoming the enabling technology for man-machine communications. The rapid development of graphics hardware and software makes its application possible. Besides building walkthroughs and landscape fly-overs with very realistic visual effects, we can recognize the trend toward industrial applications. This is because of the emerging need for tools for rapid product development. Especially in the aeronautical and automotive industries, companies have began to investigate and develop virtual reality tools for their own needs in co-operation with research organizations.

In co-operation with the Fraunhofer Institute for Computer Graphics (IGD), the Computer Graphics Center (ZGDV) in Darmstadt established the German working group on virtual reality in 1993 as a forum for information exchange between industry and research. German researchers, system developers, and industrial users have met several times in Darmstadt at the Computer Graphics Center. In these meetings they discussed the essential issues inherent in applying virtual reality to industrial applications and exchanged their latest research results and experiences.

This book summarizes the results of the meetings held in 1996. It gives a systematic view of the fundamental aspects of VR technology and its industrial applications, based on the up-to-date information provided by experts from research and industry. We would like to thank the authors for their contribution and also other participants at the meetings for the fruitful discussions. Thanks go to the colleagues who helped to organize the meetings and helped to edit this book as well.

Darmstadt, October 1997 *J.L. Encarnação*
F. Dai

List of Authors

Baacke, Peter, Dipl.-Ing.
BMW AG
D-80788 Munich, Germany

Bickel, Dieter, Dipl.-Ing.
Deneb Simulation–Software GmbH
Im Taubental 5
D-41468 Neuss, Germany

Buck, Mathias, Dipl.-Ing.
Daimler–Benz AG, Research and
Technology
P.O. Box 2360
D-89013 Ulm, Germany

Classen, Hans Josef, Dipl.-Ing.
CAE Elektronik GmbH
P.O. Box 1220
D-52001 Stolberg, Germany

Dai, Fan, Dr.-Ing., habil.
Fraunhofer Institute for Computer
Graphics
Rundeturmstr. 6
D-64283 Darmstadt, Germany

Gomes de Sa, Antonino, Dipl.-Ing.
BMW AG
D-80788 Munich, Germany

Hagemann, Franz-Michael, Dipl. -Ing.
Audi AG
D-85045 Ingolstadt, Germany

Knöpfle, Christian, Dipl.-Inform.
Fraunhofer Institute for Computer
Graphics
Rundeturmstr. 6
D-64283 Darmstadt, Germany

Kühner, Heike, Dipl.-Inform.
Computer Graphics Center (ZGDV)
Rundetrumstr. 6
D-64283 Darmstadt, Germany

Purschke, F.
Volkswagen AG, Corporate Research
D-38436 Wolfsburg, Germany

Rabätje, R.
Volkswagen AG, Corporate Research
D-38436 Wolfsburg, Germany

Schiefele, Jens, Dipl.-Inform.
Technical University Darmstadt
Flight Mechanics & Control (FMRT)
D-64287 Darmstadt, Germany

Schulze, M.
Volkswagen AG, Corporate Research
D-38436 Wolfsburg, Germany

Seibert, Frank, Dipl.-Inform.
Computer Graphics Center (ZGDV)
Rundetrumstr. 6
D-64283 Darmstadt, Germany

Starke, A.
Volkswagen AG, Corporate Research
D-38436 Wolfsburg, Germany

Symietz, M.
Volkswagen AG, Corporate Research
D-38436 Wolfsburg, Germany

Zachmann, Gabriel, Dipl.-Inform.
Frauenhofer Institut for Computer
Graphics
D-64283 Darmstadt, Germany

Zimmermann, Peter, Dipl.-Ing.
Volkswagen AG, Corporate Research
D-38436 Wolfsburg, Germany

Table of Contents

Introduction - Beyond Walkthroughs

Fan Dai
Fraunhofer-Institute for Computer Graphics, Germany

Motivation

Innovative products and shorter time-to-market are essential for the competitiveness of every company. It is therefore very important for the industry to apply new methods and tools of information technology to support the product development, production planning and other processes concerning a product. Virtual reality is one of these enabling technologies. Looking at the history of CAD/CAE techniques, we can identify the trend from pure computation to the support of creative work. In the late 1950´s, systems for finite element analysis, electronic circuit design and NC-programming were introduced to the industry. Ten years later, appeared the first systems for computer aided design with 2D and 3D wireframe graphics. Now, every CAD/CAE system has a window/menu-based interface with shaded 3D models. The enhancements on presentation quality and interactivity of these systems has led to much more efficient support of design and engineering.

Virtual reality (VR), characterized by real-time simulation and presentation of, and interaction with a virtual 3D world, offers a new quality of presentation and interaction. It has been applied to architectural design in the form of so called *walkthrough* applications. The application of virtual reality to other industrial area, e.g. mechanical engineering, is just beginning. But, as a lot of studies and sample applications from the research and industry have already shown, there is a great potential in this area (e.g. [Grebner&May95, Dai et.al.96]). It is therefore important now to develop technology and applications to meet practical industrial requirements. In the following, I will discuss the main aspects of this development briefly.

Virtual prototyping

The typical product development cycle, *concept → design → evaluation → redesign,* consists of several sub processes such as prototyping, simulation and optimization. An important way to reduce the product development cycle time is to accelerate the prototyping process. But even using the so called rapid prototyping methods, building a prototype is very time consuming and expensive. Additionally, prototypes built with generative processing methods like stereolithography do not provide sufficient material and dynamic functionality for testing. *Virtual prototyping* aims at using the CAD data of a product in combination with virtual reality tools to replace, at least to some extent, physical prototypes. The result is rapid and efficient prototyping with additional benefits: a virtual prototype – *a simulated product* – can be quickly reproduced, easy modified and transported over a network.

In the industry, a lot of design evaluations are already done electronically using simulation systems. But physical mock-ups are still used in most cases. The benefits of physical mock-ups arise from their spatial presence. Especially for conceptual design and product presentation, one can touch it, put it into the hand, and manipulate it to see if it works properly. Virtual reality techniques can be applied here to present the digital data realistically and manipulate it intuitively.

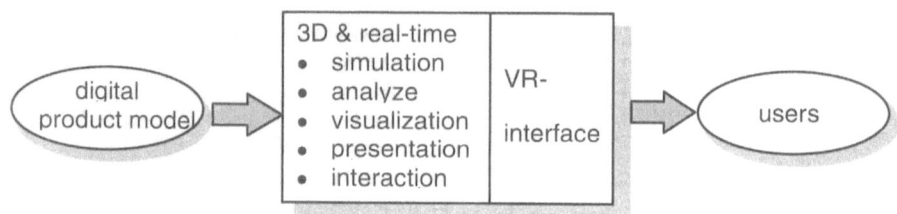

Figure 1. The idea of virtual prototyping

A system environment for virtual prototyping consists of the following components [Dai&Göbel94]: the digital product model, which integrates all information about the product and links different phases of product development; CAD and other modeling systems, which provide basic data like geometry, material, color and functional features; Simulations, which make information about the dynamics behavior and other physical characteristics available to the presentation. In this context, the virtual prototype is not only the product data itself, but also the user's imagination upon viewing and manipulating this data. The virtual reality interface enables this imagination. Advanced interaction tools allow the intuitive manipulation of the product model, and the audio-visual and haptic feedback give the user the impression of manipulating a real object.

Besides visualization of simulation data, some simulations and analysis can be done in real-time so that an investigation in the form of plug and play can be performed in the virtual world. One example of this is simulation of flexible parts. Using such built-in simulations, modeling/deforming of objects can be done simply by pressing or dragging the surface of an object. Ideally, all simulations could be run online, but this is limited by the computing power and resources. Even interactive visualization of complex data is still a problem which requires further R&D work.

Real-time 3D simulation and VR

The idea of virtual prototyping is not restricted to design evaluations. It also aims at supporting all downstream processes including production planning, training, marketing and service. As is well known, 3D simulation was applied to robot programming, factory planning and training a long time before VR was introduced to the industry. We can regard virtual reality as the continuous development of interactive, real-time 3D simulation with innovative input and output devices. Of course, not all 3D simulation applications need virtual reality interfaces, but in most cases, virtual reality interfaces will enhance the quality of presentation and interaction. The industry is becoming interested in integrating traditional 3D simulation with the new virtual reality devices and interaction methods. This integration can be done in three different ways: import the simulation results into a (native) virtual reality system and show it as an animation; integrate simulation modules into a virtual reality system; or couple the simulation system with virtual reality interfaces.

The first method is straight forward, but is limited to the viewing of pre-generated sequences. Additionally, there is often a loss of information by converting data from the simulation system into a VR system. The second and third methods try to meet the same goal from two different directions. Using an existing simulation system has the advantage that existing data can be used without conversion and remain consistent. But existing simulation systems have also their limitations because they always cover a special area. These systems have to be integrated with other systems too, especially if we want to support concurrent engineering.

Concurrent engineering

Concurrent engineering (CE) is an organizational as well as technical process to reduce the product development cycle time and to improve product quality. The essential aspect of CE is to allow experts from various areas of product development to be able to work simultaneously on the same product. In order for this to be feasible, information must be exchanged efficiently and consistently. Due to the inhomogeneous information sources, a great part of information exchange is accomplished through discussions. The European AIT-consortium (Advanced Information Technology in Design and Manufacturing) e.g. has identified a project called Digital Mock-Up (DMU) with the objective to develop the technology that allows designers, manufacturing planners and even management to work on Virtual Products / Digital Mock-Ups (DMU). They can make decisions related to the concept, design and all downstream processes. Virtual reality is one of the enabling technologies [AIT95, Dai&Reindl96].

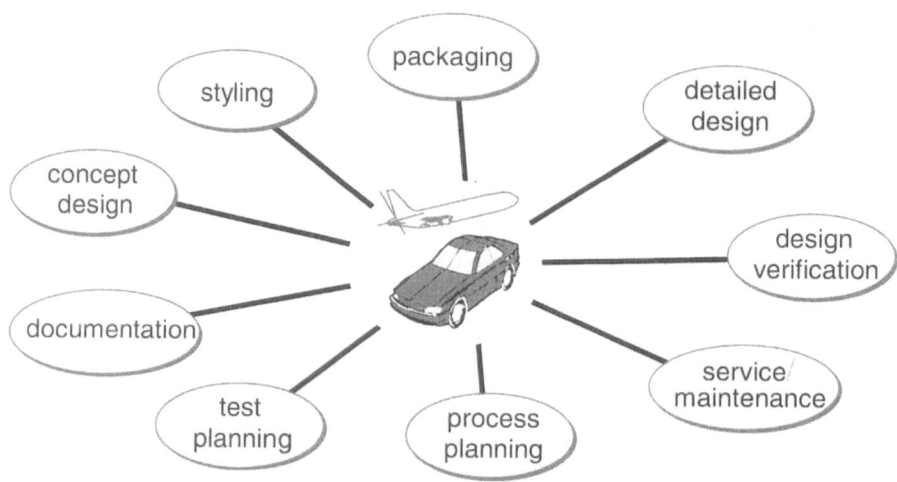

Figure 2. The vision of DMU [AIT95]

A virtual environment for concurrent engineering has to meet the following requirements:

- the information about the product must be presented in such a way that all people involved in the CE team can understand it.

- evaluations of the product design must be supported so that the CE team can make decisions.
- due to the distributed nature of CE teams, online communication over a network should be possible.

The first two requirements could be met using virtual reality techniques. In this context, virtual prototypes are used as discussion platforms for CE teams. The media of presentation and interaction is very important for concurrent engineering. A virtual environment does not necessarily use immersive displays. Dependent on the complexity and size of the interested parts, a desktop device sometimes provides better results. Therefore, the HW and SW configuration should be chosen well according to the individual application domain.

The visualization of information and the information´s relationship to the geometry are also important for CE. Both for 3D geometry and information space, navigation and manipulation should be as easy as possible. The first approachs of multimodal interaction in virtual environments are known. The combination of gesture, speech and more is the natural and intuitive way, which is especially interesting for concurrent engineering.

Supporting distributed collaborative work in virtual environments is a current research topic and is also very important. There are some basic methods and tools of CSCW (computer supported collaborative work) which can be adapted to virtual environments. Internet-based distributed systems, which are now primarily developed for online services and/or so called virtual communities have a lot in common with virtual environments for concurrent engineering. The approachs for networking, communication and avatar presentation can be adapted to distributed engineering environments too. Of course, collaborative work in a three dimensional environment using direct interaction is quite different to 2D environments. Additionally, the complexity of an interactive 3D CAD/CAE scenario requires new networking strategies.

Integrated product development

Because concurrent engineering as well as virtual prototyping cover all phases of the product development process, the virtual environment has to accomodate all kinds of product information. Virtual environments cannot replace CAD/CAE systems. They simply provide an additional platform with different, new functionality. Detailed design and analysis will be done mostly with „native" CAD/CAE systems. In an environment for integrated product development, virtual reality will be used in several areas with different functionality, in parallel with CAD/CAE systems. Therefore, a seamless link between virtual environments and other systems is very important. This means not only an one way interface from CAD/CAE to VR, but also in the other direction.

Product data management and engineering data management (PDM/EDM) systems play the central role in integrated product development. The link between virtual environments and PDM/EDM is therefore important for future development of VR applications, e.g., to get product structure and configuration into the virtual environment, and to put review results back into the database. These aspects are being addressed by industry and research.

Attempts to standardization, both for CAD/CAE (STEP [ISO94]) and for virtual environments (VRML [VRML96]) must also be mentioned. These developments are very important for our goal of virtual environments for engineering applications. Standardization of data exchange provides the basis for the integration of different systems, including virtual reality tools. A STEP-oriented data format for virtual environments, for example, could be developed as the basis for the integration via PDM/EDM systems.

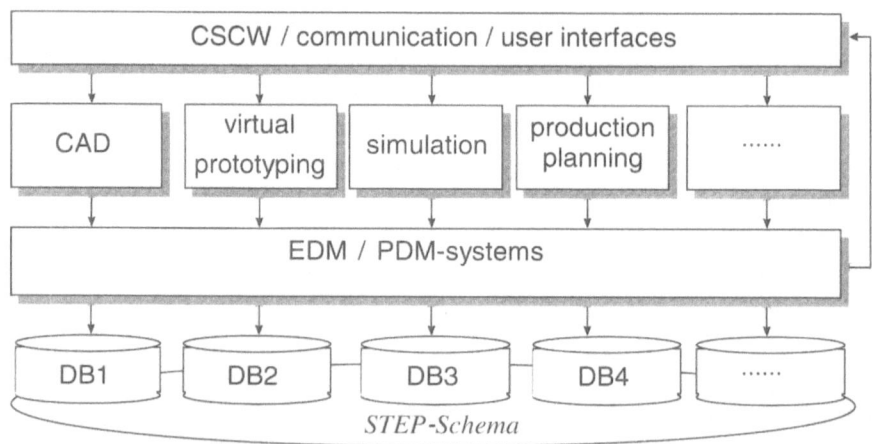

Figure 3. Integrated product development

Current situation and trends

For a few years now, several commercial VR systems have been available. Included in this group are WorldToolKit from Sens8 and dVISE from Devision, Inc. With these systems, one can create virtual environments with textured 3D objects, virtual buttons, etc. Using the authoring tools and the programming interfaces provided, one can build individual applications. Of course, these systems have only limited functionality, mainly for visual experience of 3D worlds such as walkthroughs of virtual rooms. Research systems offer more functionality in the areas of collision detection, online analysis/simulation and advanced visualization and interaction methods as well as the support of new input/output devices. These systems are still under development to meet the requirements of industrial applications.

Existing methods for interacting in virtual environments are mainly based on virtual menus/buttons and gestures. Methods for direct, intuitive manipulation of virtual objects are under development. This includes the development of online simulation methods for collision, contacts and deformations.

One of the problems common to all these systems is the link to CAD/CAE systems. Currently, there is no common data format for CAD/CAE and VR systems. Each system uses a special data format. Virtual environments have different requirements than CAD/CAE systems: visual realism and interactivity. Therefore, the data formats of VR systems are quite different from those of CAD/CAE systems. To be able to import/export at least the geometrical data, IGES and VRML are often used as the data interface, besides the direct conversion between native formats. Preparation tools are developed to do corrections on the geometry data, add visual attributes and add functional information.

VR systems have now left their beginner's status, which was characterized by walkthroughs and games. The development of more intelligent tools for scenarios with more functionality is important for engineering applications. It is an interdisciplinary work, which requires collaborative R&D between information scientists and mechanical engineers. As mentioned above, there are already sample applications which demonstrate the application potentials of VR and show some possible ways. There are also a lot of research projects in progress. Looking back on the rapid development of graphics HW and SW, and compared with the success of CAD/CAE applications, we are sure that we will be successful in the development of virtual environments for engineering, too.

Contents of this book

There are many publications now containing detailed explanations of the fundamental methods of rendering and the corresponding input and output devices. This book does not attempt to give an introduction to the basics of virtual reality technology, but rather to present some new techniques and the applications of existing technology.

Following this introduction, the reader will find three parts. In the first part, *VR Techniques*, three contributions describe some essential aspects of VR. Zachmann explains the general structure of a VR system and concentrates on the techniques needed for industrial applications. Buck describes his approach on direct object manipulation based on the simulation of rigid body contacts and dynamics with some examples realized at Daimler-Benz. Kühner presents a new method for video-based input with her prototype implementation using the ViP (virtual plane) as a hardware platform.

In Part 2, *From CAD to VR*, we present two approachs to data preparation. Hagemann describes an approach using a native CAD system (CATIA), which has the advantage that the semantic information are available. But today, in most cases, we do not have the possibility to modify a CAD system. We have to make use of the tessellated geometry data. The data preparation takes place on a polygon basis. Knöpfle and Schiefele describe their approach, which has been used in recent industrial projects.

Part 3, *Applications*, finally contains four contributions from the industry. Bickel presents his view on virtual reality for industrial applications based on several examples which are realized with a realtime 3D simulation system. Puschke et al. show us the vision of and some existing and future applications at Volkswagen. Gomes de Sa and Baake describe their experience in the realization of a concrete assembly study application using a commercial VR system. Finally, Claßen presents some applications from the space industry.

This book certainly does not cover all aspects related to VR for industrial applications. But the readers may find interesting examples and get important stimulation for their own development of VR applications. The goal of this book is to share the valuable experiences of the experts with all researchers, developers and users of virtual reality techniques.

References

[AIT95] AIT-consourtium: Project report of the pilot phase. June 1995, AIT project office at Daimler-Benz AG, Germany.

[Dai&Reindl 96] Dai F., Reindl P., Enabling Digital Mock-Up with Virtual Reality Techniques. Proceedings of ASME Design Technical Conferences, September 1996, Irvine, CA USA.

[Dai et al. 96] Dai F., Felger W., Frühauf Th., Göbel M., Reiners D., Zachmann G., Virtual Prototyping Examples for Automotive Industries. Virtual Reality World '96. Stuttgart, Germany, Feb. 13. - 15. 1996.

[Dai&Göbel 94] Dai F., Göbel M., Virtual Prototyping - an Approach using VR-techniques. In: Proceedings of the 14th ASME Int. Computers in Engineering Conference, Minneapolis, Minnesota, September 11-14, 1994.

[Grebner&May 95] Grebner K., May F. (eds), Application of Virtual Reality Techniques in the Industry - selected Examples. Conference documentation of Virtual Reality World 1995, pp.451-468.

[ISO 94] ISO/IS 10303-1: Industrial automation systems and integration - Product data representation and exchange - Part 1: Overview and fundamental principles . International Organization for Standardization; Geneve (Switzerland); 1994.

[VRML96] The VRML consortium, The virtual reality modelling language, Specification version 2.0, August 1996. (http://www.vrml.org)

Part I

VR Techniques

1 VR-Techniques for Industrial Applications

Gabriel Zachmann
Fraunhofer Institute for Computer Graphics, Germany

Virtual Prototyping (VP) is, so far, the most challenging class of applications for virtual reality (VR). A VR system suitable for VP must be able to handle very large geometric complexities which, in general, cannot be reduced significantly by common rendering techniques such as texturing or level-of-detail. Furthermore, VP is the most „interactive" type of applications for VR compared to other areas such as architecture, medicine, or entertainment. This is due to the simulation of rather complex tasks involving many objects whose behavior and properties should be imitated as closely as possible. Finally, in order to sustain continuously a feeling of presence and immersion, the VP system must be able to achieve a frame rate of no less than 15 frames per second at any time.

This chapter will deal with the issue of interaction in complex virtual environments (VEs) for VP. First, we will review briefly the novel I/O devices currently being used in VR, and classify VEs with respect to the devices being used and with respect to the real environment being simulated. Then, several interaction techniques will be discussed, both for VEs in general and for VP in particular. The last two sections give a more technical view on the issue of how to describe VEs and on the overall architecture of VR systems.

1.1 Characterization of VEs

Virtual environments can be distinguished by their relation to a real environment.
- The VE is actually a *projection* of some real environment. That real environment might be of very different scale [28] or at some distance from the user [5]. The latter is usually described by the term *telepresence*.
- The VE does *not exist* but is otherwise (fairly) realistic. It might have existed in the past, or it might exist in the future (which is actually the purpose of VP).

- The VE is quite unreal. This is commonly the case in entertainment which strives to provide the participants with an exciting and exotic world.

There are, of course, intermediate or mixed forms. An example of this is „augmented reality" [8]: the user sees his real environment through special glasses which allow the superposition of computer generated images. Thus, the image of the real world can be augmented by pictures, signs, pictograms, hints, instructions, etc. However, interaction is usually restricted to the real environment.

Different VEs (as distinguished above) require different kinds of interaction:

- the user manipulates *existing*, real objects. Examples are: repair of a satellite by means of a vehicle armed with tools; steering of a vehicle where humans cannot enter (e.g., the moon, radioactively polluted terrain, etc.); exploration and modification of the structure of materials at atomic scale [28]
- the user manipulates *non-existent* objects. Examples might be: examination and modification of the interior design of a building which is still under planning; visualization of fabrication processes, such that deficiencies or dangers for human operators can be detected at an early stage; simulation of surgical operations.

Another classification scheme can be based on the amount of *distribution*. The simplest VEs are local, single-user. Distributed VEs might still be single-user, but the application is distributed among several machines or processors. In *multi-user* VEs, several users share the same experience while being at (possibly) remote places.

1.2 Techniques to achieve immersion

Virtual environments can be classified according to several orthogonal criteria: the real environment being simulated, the amount of *immersion* and *presence* they offer (see Figure 1.1), and the degree of distribution.

1.2.1 Immersion

A key feature of VR technology is *immersion*. This term defines the feeling of a VR user, that his virtual environment is real[1].

A high degree of immersion is equivalent to a realistic or „believable" virtual environment. In the following we will briefly describe several effects which detract from the experience of immersion (ordered by significance), and how they can be avoided:

1. Psychological experiments indiciate that the most important effect is *feedback lag*. There are mainly three factors which contribute to this lag:
   - Rendering time.
   - Tracking systems usually produce delayed data. The more sensors they have to track, the longer the delay. In addition, filtering introduces lag. Attempts have been made to overcome this problem by trying to predict the position/orientation [17].
   - Other computations like collision detection or simulation of physics.
2. *Narrow field-of-views* is a rather severe shortcoming which makes the user feel as if he looked through a tunnel into the world, especially, if he uses a head mounted display. To overcome this, HMDs usually have got some kind of wide-angle optics. Large projection screens can alleviate the „tunnel" effect of monitors; a very large field-of-view is achieved by a surround-screen projection (cave) as developed by [6]
3. A *monoscopic* view deprives the user of ability to estimate distances in the depth range from 20cm to about 5m. A stereoscopic view can be obtained quite easily by simply rendering the same scene twice with slightly shifted viewpoints (stereoscopic rendering). One should use the parallel-lens algorithm to achieve a good stereoscopic effect [18, 1, 27]. Unfortunately, the conflict between the computer generated disparity in the user's eyes and the eye's accommodation to the screen surface (or the HMD's virtual display distance) will still remain. From our experience, this is not a problem, although the discrepancy is not realistic.
4. *Low display resolution*, from our experience, is the least significant factor concerning immersion. First generation HMDs based on LCDs have very low resolution which is made worse because of the wide-angle optics. CRT based HMDs offer full resolution, but are heavier.

Since our visual perception is our primary sense, research focuses very much on fast, realistic, high-resolution rendering. However, other senses

---

[1] More precisely, we define complete *immersion* analogously to Turing's definition of (artificial) intelligence: if the user cannot tell which reality is „real", and which one is „virtual", then the computer generated one is totally *immersive*.

must also be stimulated to achieve full immersion; among them are audible feedback, tactile, and force feedback.

1.2.2 Presence

While immersion is an objective measure, presence is the subjective sense of *being in* the VE [26]. Presence requires a self-representation in the VE–a virtual body (often, only the hand is represented in the VE, since the whole body is not tracked). It requires also that the participant identify with the virtual body, that its movements are his/her movements.

1.3 Devices

Virtual Reality offers a novel human-computer interface by utilizing novel I/O devices. The output devices commonly used and their determining characteristics are:

Display	Characteristics
HMD, Boom Cave, Stereo-Projection Workbench Monitor	Resolution Color / Monochrome Field-of-View Contrast and Brightness Distortion

Very important to most VR applications is the tracking device which measures the head's, hand's, or the complete body's position and orientation. Several technologies and characteristics are:

Tracking technology	Characteristics
magnetic optical mechanical other (ultra-sonic, inertial, gyroscopic)	volume noise and distortion latency accuracy, occlusion

Other input devices are:

Device	Characteristics
glove desktop positioning device tetherless positioning device other (buttons, microphone, ...)	dimension accuracy practicality

The amount of *immersion* and *presence* is determined, among others, by the I/O devices being used [11]: the ultimate immersion can be achieved by head-mounted displays, followed by arm-mounted (Boom), several cave variants through screen-based (Workbench or just stereo-projection) systems. Unfortunately, the degree of presence is (almost) reciprocal to the degree of immersion: with a workbench, we do have full presence, but almost no immersion. In a cave, we have optimal presence because the real body is actually *inside* the VE, but we cannot achieve full immersion. With an HMD we have full immersion but almost no presence.

Using output devices for other human senses such as hearing or feeling greatly increases the effect of immersion.

The situation is quite similar on the input side: all input devices so far offer different advantages and disadvantages. While one application might demand a fully immersive VE, another one might be best implemented by stereo-projection and spacemouse.

The result of this observation is that there cannot be a perfect, general input or output device. Instead, the appropriate devices are determined by the application. For assembly simulations it is important to achieve a very high level of immersion in yet to be built „environments". Style and desgin evaluation doesn't necessarily need perfect immersion, but it does need high quality images and a good sense of presence. Entertainment needs a high degree of immersion in unreal environments which will never exist. Augmented reality usually provides a very high level of presence in the existing environment, while there is not much of a virtual environment at all. Figure 1.1 gives a graphical depiction of our classification of some typical areas of VR applications.

As can be seen from the figure, VP is an area which has a very broad range of requirements: on one hand, there are applications more biased towards functional simulation, and thus requiring a high degree of immer-

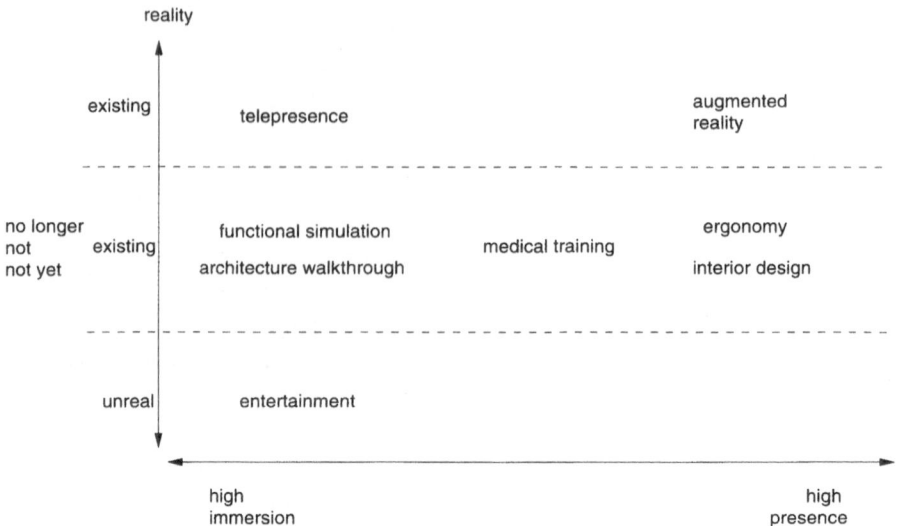

Figure 1.1. This graph depicts the classification of some typical applications of VR with respect to the „reality" of the environment and with respect to the immersion/presence required.

sion; on the other hand, there are applications such as ergonomy studies or interior design which require a high degree of presence.

1.3.1 Distortion correction of magnetic fields

Electro-magnetic trackers have become the most wide-spread devices used in today's virtual reality systems. They are used to track the position and orientation of a user's hands and head, or to track instruments such as an endoscope or scissors. They're also being deployed in real-time motion capture systems to track a set of key points and joints of the human body. Commercial optical tracking systems are becoming more mature; however, they're still much more expensive than electro-magnetic systems, but are not yet quite as robust in terms of drop-outs.

Unfortunately, there is one big disadvantage of electro-magnetic trackers: the electro-magnetic field itself, which gets distorted by many kinds of metal. Usually, it is impossible to banish all metal from the sphere of influence of the transmitter emitting the electro-magnetic field, especially when using a long-range transmitter: monitors contain coils, walls, ceiling, and floors of the building contain metal trellises and struts, chairs and tables have metal frames, etc. While tracking systems using direct current-seem to be somewhat less susceptible to distortion by metal than alternating current systems, all ferro-magnetic metal will still influence the field generated by the transmitter to some degree.

A distortion of the magnetic field directly results in mismatches between the tracking sensor's true position (and orientation) and the position (orientation) as reported by the tracking system. Depending on the application and the set-up, mismatches between the user's eye position (the *real viewpoint*) and the virtual camera's position (the virtual viewpoint) impair more or less the usability of VR. For example, in assembly tasks or serviceability investigations, fine, precise, and true positioning is very important [7]. In a 2.4m cave or at a workbench, a discrepancy of 7 cm (3 in) between the real viewpoint and the virtual viewpoint leads to noticeable distortions of the image[2], which is, of course, not acceptable to stylists and designers. For instance, straight edges of objects spanning 2 walls appear to have an angle (Figure 1.2), and when the viewer goes closer to a wall, objects „behind" the wall seem to recede or approach (see [14] for a description of some other effects). Mismatches are most fatal for Augmented Reality with head-tracking in which virtual objects need to be positioned exactly relative to real objects [22].

We have developed an algorithm which overcomes these adverse effects [34]. It is very fast (1-2 msec), so no additional latency is introduced into the VR system. With our algorithm, the error of the corrected points from their true positions does not depend on the distance to the transmitter, nor does it depend on the amount of local „warp" of the magnetic field.

1.4 Interaction techniques

By interaction we mean any actions of the user aiming at modification or probing the virtual environment. In order to achieve a good degree of immersion, it is highly desirable to develop interaction techniques which are as intuitive as possible. Conventional interaction devices (keyboard, mouse, tablet, etc.) are not fit for natural interaction with most VR applications. One of the more intuitive ways is the „virtual trackball" or the „rolling trackball", which both utilize the mouse [16, p.51 ff.].

---

[2] This is just a rule of thumb, of course. The threshold at which a discrepancy between the real and the virtual viewpoint is noticeable depends on many variables: expertise, distance from the cave wall or projection screen, size of the cave, etc.

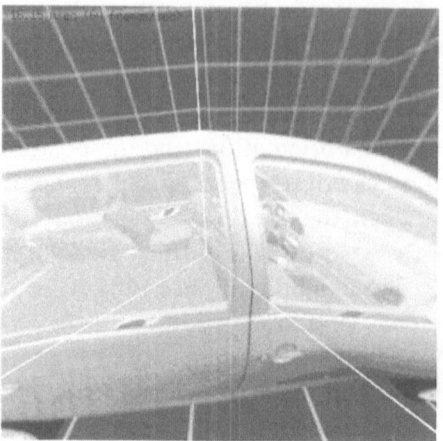 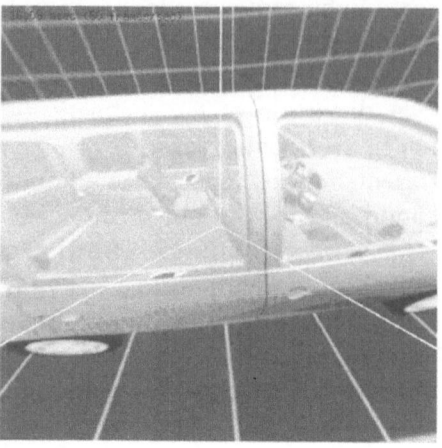

Figure 1.2. Without correction of the magnetic field tracking, an off-center viewer in a cave will see a distorted image. The effect is even worse when the viewer moves, because objects seem to move, too.

Figure 1.3. With our method, the perspective is always correct, even when the viewer moves. Data courtesy of Volkswagen AG.

The shortcoming of all of the above mentioned devices is their low number of input dimensions (at most 2). However, new devices like SpaceMouse, DataGlove, tracking systems, Boom, Cricket, etc., provide 6 and more dimensions. These allow highly efficient, natural interaction techniques. Some of them will be described in the following.

1.4.1 Navigation

By navigation we mean all forms of controlling the viewpoint in a virtual environment, or steering of a real exploration device (e.g., the repair robot or the microscope needle). Navigation is probably the simplest form of interaction, which can be found in all VR applications.

Virtually all navigation techniques can be deduced from a single model, which assumes the virtual camera mounted on a virtual cart, also sometimes referred to as *flying carpet* model (see Figure 1.4). See also [32, 25, 9].

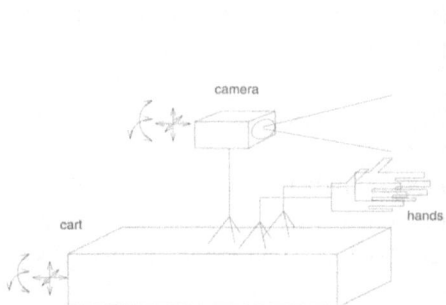
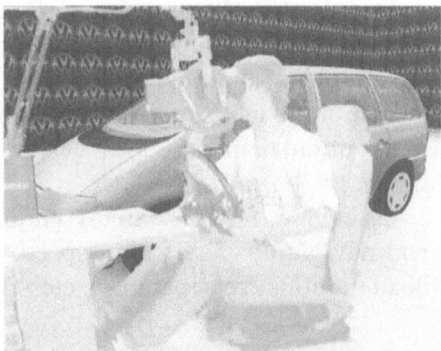

Figure 1.4. All navigation modes can be deduced from the *flying carpet* model. Not all modes utilize all of the „devices" shown.

Figure 1.5. With a Boom for interior design, the eyeball-in-hand navigation technique is used (see section Color Plates).

- In *point-and-fly* the user moves the cart by pointing in the desired direction with the navigation device (e.g., glove or cricket) and making a certain gesture or pressing a certain button. If a glove is being used, the speed of the motion can be controlled by the flexion value. If head tracking is enabled, the camera will be controlled by the head tracker. This navigation technique is the one most widely used. [19] have suggested a more sophisticated point-and-fly mode: the user points at the desired object and the VR system computes a „swerved" path which will place the user eventually in front of the object. Again, the speed can be controlled. Sometimes it is desirable to constrain the cart at a certain height, for example at eye level above the virtual ground, while the user can still move around.
- *eyeball-in-hand*: this paradigm is implemented by feeding the tracking system's output (e.g., the position of an electro-magnetic sensor or a Boom), directly to the viewpoint and the viewing direction, while the cart remains fixed. This technique is most appropriate for close examination of single objects from different viewpoints, e.g., interior design (see Figure 1.5).
- *scene-in-hand*: this is the complementary technique to eyeball-in-hand. Sometimes this can be quite useful for orientation or coarse object placement [23].
- Sometimes it is desirable to be able to control the viewpoint „without hands". In that case, *speech recognition* can be used in order to move the cart by uttering simple commands such as „turn left", „stop", etc. This has become feasible with today's user-independent speech recognition systems and fast processors.

In order to attain maximum flexibility, it is highly desirable that all of the above mentioned navigation modes can be mapped to all possible configurations of input devices. While there are certain combinations of navigation mode and input device configuration which will be utilized much more often than others, a general mapping scheme comes in handy at times.

Of course, there are many parameters which affect user representation and navigation: navigation speed, size and offset of the hand, scaling of head motion, eye separation, etc. These have to be adjusted for every VE.

1.4.2 Gesture recognition

Usually, static gestures like „fist" or „hitch-hike" are used to trigger actions. Gestures can be augmented by taking also the orientation of the glove's tracker into account, i.e., a hitch-hike gesture pointing up is different from a hitch-hike pointing down (for example, to make an elevator go up or down). There is also research going on to recognize dynamic gestures which consist of a continuous sequence of static gestures and tracker positions.

There is a little bit of confusion about terminology here. Sometimes, static gestures are called *postures* and dynamic gestures are just called *gestures*. We will call gestures plus orientation postures.

Gestures can be defined as ellipsoids in $\mathbf{R}^d$ (where d=10 or d=20, typically); then, they can be recognized by a simple point-in-ellipsoid test. Of course, other norms can be utilized as well, for example the l^∞ -norm is computationally much more efficient.

Another, more robust approach exploits the fact that almost all gestures are located near the „border" of $[0,256]^{20}$, i.e., their flex values are (almost) maximal or minimal. So, $\mathbf{R}^d$ can be subdivided into certain half-spaces, quarter-spaces, etc.

Other approaches are back-propagation networks (perceptrons) [30] and hidden Markov models [21]. Still, with all algorithms, glove calibration for each user is necessary.

Gestures are very well suited to trigger simple actions, like navigation or display of a menu. However, experience has shown that VR systems should not be over-loaded with gesture driven interaction [24]. A casual user will confuse gestures, and the every-day user will find them rather unnatural. In fact, a set of gestures which trigger specific actions can be considered another type of (invisible) menu, which is no more realistic or natural than well-known 2D desktop menus. Other techniques such as speech recognition or natural object behavior should be utilized whenever there is no special reason to use gestures.

1.4.3 3D menus

Menus provide a 1-of-n choice. 3D menus are a straight-forward extension of the well-known 2D menus [15]. Usually their appearance is triggered by a gesture or a button. Several possibilities exist in order to select a menu item with the pointing device (glove, flying joystick, etc.): shooting a ray through the index finger (if a glove is being used), shooting a ray from the eye through the index finger, or actually touching the 3D button with the finger.

However, we feel that 3D menus should be avoided. They are a relic of 2D desktop interaction and there are usually more efficient ways to interact in 3D. Furthermore, almost all 1-of-n choices can be done much more efficiently by speech recognition.

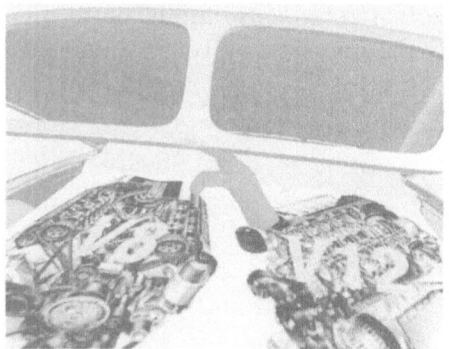
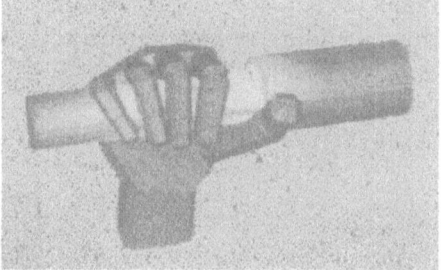

Figure 1.6. Objects should behave naturally, so interaction with them can be natural. Here, the hood of the car can be opened by just lifting it (see section Color Plates).

Figure 1.7. When objects should be grabbed, interaction with virtual environments can actually be more efficient than in the real world.

1.4.4 Natural object behavior

The goal of this technique is to avoid burdening the user with acquiring any skills other than the ones used in every-day life, i.e., objects in the virtual world should behave just like they do in the real world. However, to achieve such realism, great computational resources are needed, both in terms of CPU power and in terms of clever algorithms.

Grabbing, pushing, or pulling an object are every-day tasks. As in the real world, a user should be able to do this in the virtual environment by just pushing or pulling it with a virtual hand (see Figure 1.7, Figure 1.6).

Although it is highly desirable that objects behave naturally, we feel that the designer of a virtual environment has to decide on a case-by-case basis how close to reality certain interactions should be. For example, for many tasks, it is perfectly sufficient to let the user grab an object by just touching it while making a fist gesture. However, for other scenarios, e.g design evaluation, the grabbing should be modeled as close to reality as possible.

Another property of realistic object behavior is that they do not penetrate each other. The basis of almost all natural object behavior is real-time collision detection for complex objects [12]. However, if force feedback is not available, collision response must be modeled carefully to facilitate efficient interaction. For example, it might be very hard to place an object within a dense environment, if colliding objects stick to each other [29]. Instead, a more suitable collision response might be to highlight colliding objects and accompany the collision by some sound feedback (see Figure 1.8).

Figure 1.8. Collision response while moving an object must not interfere with smooth interaction. Here, collision response was chosen to be highlighting of colliding objects by wireframe and via sound feedback (see section Color Plates).

Figure 1.9. Particle sources can be placed in a virtual environment interactively, while the particle tracing module computes particle paths on-line. So, the simulation of a flow field can be visualized within the environment itself, which is the inside of a car in this case (see section Color Plates).

1.4.5 Interaction for scientific visualization

This is a rather new field in the realm of virtual environments. By using 6D (or more) devices, new and potentially much more efficient techniques for interaction with simulation results can be devised.

We have ported several visualization techniques for flow fields to virtual environments. Common techniques are particle tracing, stream lines, and streak lines. For each of them, sources within the field have to be placed. Using interaction techniques for virtual environments, it is very easy to place these: the user just grabs the object which represents the source and places it somewhere else in space. Or, even more efficient, the fingers of the virtual hand itself become sources (see Figure 1.9).

1.4.6 Cooperation in VEs

So far, computer-supported cooperative work (CSCW) does not play a major role in today's product design. As companies operate more and more globally, however, design teams will get more and more distributed. Eventually, there will be a need for CSCW techniques in VP systems, so that several designers or stylists can discuss a new digital prototype while being at remote sites, possibly located on different continents. Therefore, a VR system should be able to allow multiple users to interact with each other and with the same VE (see Figure 1.10 for an example).

From a classification point of view, it does not make a big difference if there are multiple participants or just one. However, from a technical point of view, it *does* make quite a difference. When dealing with multiple participants, many new technical issues arise: for instance, the time lag introduced by networks should be neutralized; appropriate representations of the other users should be created.

A great challenge is *floor control*, i.e., mutually exclusive access to objects, in real-time across networks. Intertwined with that is the problem of when access should be mutual exclusive: sometimes participants *do* want to access an object simultaneously (e.g. for repair), sometimes it is not necessary (because the attributes being altered are „orthogonal"). Several methods have been devised to insure consistency:

- *Ownership token passing* [31]. Smooth dead-reckoning is done for local ghosts. The owner of the token of an objects broadcasts updates for that object. When and how the token for an object will be passed is up to the implementation and might be determined by the application.

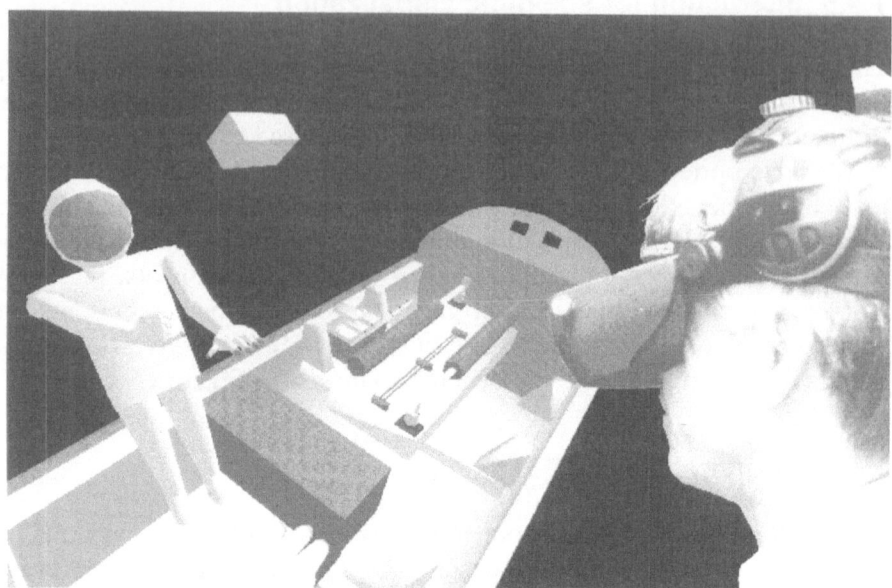

Figure 1.10. In a shared virtual environment, two astronauts perform a repair task on the Hubble telescope while being at different locations (Germany and Texas). Scenarios like this can be used for training or simultaneous engineering.

- *Read/write* permissions à la Unix. Objects can be marked writable for a specific group of users. Every process which changes an attribute of an object and has write permission, broadcasts updates.
- *Distributed locks.* When a process wants to change an object, it will first lock it's local database, then request a distributed lock on that object. With this approach it might be difficult to assert responsiveness at all times. In order to avoid delays in the simulation loop, an application could try to look ahead and request a *lock in advance* if it can be foreseen that a user will probably try to execute an action which requires a particular lock.

The methods above can be combined – however, whatever measures we take, we must make sure that the overall system is *responsive* at all times.

1.5 Description of VEs

Creating virtual worlds is still a cumbersome and tedious process. Below, we describe a framework which facilitates creating virtual environments. VE „authors" should be allowed to experiment and play interactively with their „worlds". Since this requires very low turn-around times, any compilation or re-linking steps should be avoided. Also, authors should not need to learn a full-powered programming language. A very simple, yet powerful script language will be proposed, which meets almost all needs of VE creators [33] (we will not, however, discuss any syntactical issues, since they can be found in the reference).

In order to achieve these goals, we identify a set of basic and generic user-object and object-object interactions which, experience has taught us, are needed in most applications.

For specification of a virtual world, there are, at least, two contrary approaches:

- The *event based* approach is to write a *story-board*, i.e., the creator specifies which action/interaction happens with a certain event. A story-driven world usually has several „phases", so we want a certain interaction option to be available only at that stage of the application, and others at another stage.
- The *behavior based* approach is to specify a set of *autonomous objects* or *agents*, which are equipped with receptors and react to certain inputs to those receptors (see, for example, [4]). So, overstating a little, we create a bunch of „creatures", throw them into our world, and see what happens.

In the long term, you probably want to be able to use both methods to create a virtual world.

Here, we will focus on the *event based* approach. The language for specifying those worlds will be very simple for several reasons: VE authors „just want to make this and that happen", they don't want to learn Python or C++. Moreover, it is much easier to write a true graphical user interface for a simple language than for a full-powered programming language.

All concepts being developed here have been inspired and driven by concrete demands during recent projects. Most of them have been implemented in an *interaction-module*, which is part of our whole VR system.

1.5.1 The action-event paradigm

A virtual world is specified by a set of *static* configurations (geometry, module parameters, navigation modes, etc.) and a set of *dynamic* configurations. Dynamic configurations are object properties, user-object interaction, action dependencies, or autonomous behavior.

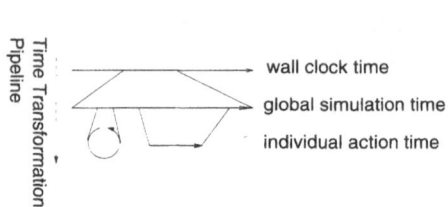

Figure 1.11. A simulation of virtual environments must maintain several time „variables". Any action can have its own *action time* variable, which is derived from a global *simulation time*, which in turn is derived from wall clock time. There is a small set of actions which allow the simulation to set/change each time transformation individually.

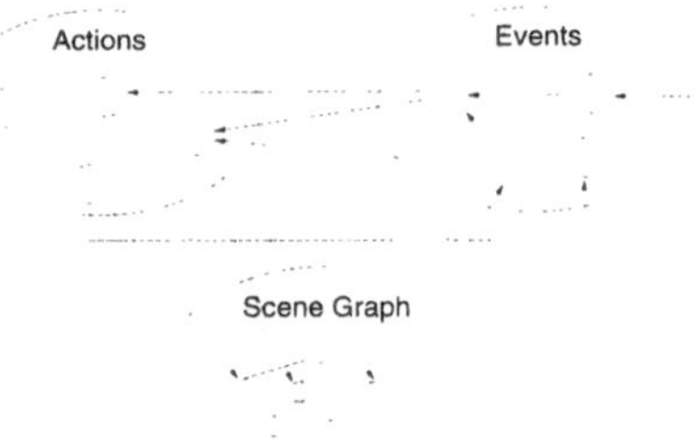

Figure 1.12. The *AEO triad*. Anything that can „happen" in a virtual environment is represented by an action. Any action can be triggered by one or more events, which will get input from physical devices, the scene graph, or other actions. Note that actions are not „tied-in" with graphical objects, and that events are objects in their own (object-oriented) right.

The basic idea of dynamic configurations is that certain *events* trigger certain *actions*, *properties*, or *behavior*; e.g., when the user touches a virtual button, a light will be switched on, or, when a certain time is reached an object will start to move. Consequently, the basic building blocks of our virtual worlds are *actions*, *events*, and *graphical objects* – the *AEO triad*[3] (see Figure 1.12).

Unlike other systems [13, 2, 10], our actions are *not* part of an object's attributes (in fact, one action can operate on many objects at the same time).

In order to be most flexible, the action-event paradigm must satisfy the following requirements:

1. Any action can be triggered by any event.

---

[3] In object-oriented programming parlance, actions, events, as well as graphical objects are *objects*. However, in the following we will use the term *object* only for graphical objects.

2. Several events can trigger the same action. An event can trigger several actions simultaneously (many-to-many mapping).
3. Events can be combined by boolean expressions.
4. Events can be configured such that they start or stop an action when a certain condition holds for its input (positive/negative edge, etc.)
5. The status of an action can be the input of another event.

We do not need any special constructs (as in [20]) in order to realize *temporal operators*. Parallel execution of several actions can be achieved trivially, since one event can trigger many actions. Should those actions be triggered by different events, we can couple them via another event. Sequential execution can be achieved by connecting the two actions by an event which starts the second action when the first one finishes. Similarly, actions can be coupled (start-to-start or start-to-stop) with a delay.

1.5.2 Time

Many actions (besides navigation, simulation, and visualization) depend on time in some way. For example, an animation or sound sample is to be played back from simulation time t_1 through t_2, no matter how much computation has to be done or how fast rendering is.

We maintain a global *simulation time*, which is derived from wall-clock time. The transformation from wall-clock time to simulation time can be modified via actions (to go to slow-motion, for example, or to do a time „jump").

Furthermore, we keep an unlimited number of time variables. The value of each time variable is derived from the global simulation time by an individual transformation which can be modified by actions as well (see Figure 1.11).

Those times can be used as inputs to events, or to drive simulations or animations. Thus, time can even be used to create completely „time-coded" parts of a virtual reality show.

1.5.3 Events

Events are probably the most important part for our world description – they can be considered the „sensory equipment" of the actions and objects. They have the form

event-name: trigger-behavior input parameters

where *event-name* is for further reference in the script. When an event „triggers" it sends a certain message to the associated action(s), usually „switch on" or „off".

It is important to serve a broad variety of inputs (see below), but also to provide all possible trigger behaviors. A trigger behavior specifies when and how a change on the „input" side actually causes an action to be executed. Let us consider first the simple example of an animation and a keyboard button:

> *animation on as long as button is down,*
> *animation switch on whenever button is pressed down,*
> *animation switch on whenever button is released,*
> *animation change status whenever button is pressed down,*

These are just a few possibilities of input→action trigger-behavior. The complete syntax of trigger behaviors can be found in [33].

It would be possible to have the world builder „program" the trigger-behavior by using a (quite simple) finite state machine (as in dVS, for instance [10]). However, we feel that this would be too cumbersome, since those trigger behaviors are needed very frequently.

In addition to the basic events, events can be combined by logical expressions. This yields a directed „event graph". This graph is not necessarily acyclic.

Our experience shows that it is necessary to be able to activate and deactivate actions. This is needed, for example, to enable unmounting of a part only after some screws have been removed. This is done via a certain action, which (de-)activates other actions.

A collection of event inputs.

Physical input includes all kinds of buttons (keyboard, mouse, spacemouse, boom), flex and tracker values, gestures, postures (gesture plus orientation of the hand), voice input (keyword spotting, enhanced by a simple regular grammar, which can tolerate a certain (user-specified) amount of „noise" words).

Geometric events are triggered by some geometric condition. Among them are virtual buttons, virtual menus, portals, and collisions.

Any action's status (*on* or *off*) can trigger an event. Some actions have an action-specific status, which can be used also.

All time variables (see above) can be the input of an event. This allows for one-shot or cyclical triggering of actions.

Sometimes we need to „monitor" certain object attributes and issue an action when they change, while we don't care which action (or even other module) changed them. Attributes are not only graphical attributes (transformation, material, wireframe, etc.), but also „interaction" attributes, such as „grabbed", „falling", „constrained", etc. Object attributes might be set by our interaction module itself (by possibly many different actions), or by other modules.

1.5.4 Actions

Actions are the building blocks for „life" in a virtual environment. Anything that happens, as well as any object properties, are specified through actions. The set of actions should be very generic, but not too low level. If they are too low level, then building VE with them is probably no more efficient that using a programming language. If they are too high level, then they are probably too specialized and cannot be used in a broad variety of applications.

Actions are usually of the form

action-name: function objects parameters options

All actions should be made as general as possible, so it should always be possible to specify a *list of objects* (instead of only one). Also, objects can have any type, whenever sensible (e.g., assembly, geometry, light, or viewpoint node). The *action-name* is for later reference in the script.

There are several cases where inconsistency has to be dealt with in a VR system. One such case arises when several actions transform the same object. For example, we can grab an object with our left hand while we stretch and shrink it with the other hand. The problem also arises, when an action takes over (e.g., we scale an object after we have grabbed and moved it). However, by implementing a standardized way of object transformation, this problem can be solved.

Another inconsistency arises when we use levels-of-detail (LODs) in the scene graph. Since any object can be a LOD node or a level of a LOD, any action should transparently apply changes to all levels, so that the author of the virtual world doesn't have to bother or know whether or not an object name denotes an LOD node (or one of its children).

A collection of actions

During our past projects, the set of actions listed below briefly has proven to be quite generic.

The scene graph can be changed by the actions load, save, delete, copy, create (box, ellipsoid, etc.), and attach (changes scene hierarchy by rearranging subtrees).

Some actions to change object attributes are switch, wireframe, rotate, translate, scale (set a transformation or add/multiply to it). Others change material attributes, such as color, transparency, or texture.

The „grab" action first makes an object „grabbable". Then, as soon as the hand touches it, it will be attached to the hand. Of course, this action allows grabbing a list of sub-trees of the scene graph (e.g., move a table when you grab its leg).

With the „stretch" action we can scale an object (or sub-tree).

A great deal of „life" in a virtual world can be created by all kinds of animations of attributes. Our animation actions include playback of transformations, visibility, transparency (for fading), and color from a file. The file format is flexible so that several objects and/or several attributes can be animated simultaneously. Animations can be time-tagged, or just be played back with a certain, possibly non-integer, speed. Animations can be *absolute* or *relative* which just *adds* to the current attribute(s). This allows, for example, simple autonomous object locomotion which is independent of the current position.

As described above, there are actions to set or change the time transformation for the time variables.

Occasionally we want to constrain the movement of an object. It is important to be able to switch constraints on and off at any time, which can be done by a class of constraint actions. Several constraints on transformations of objects, including the viewpoint, have proven useful:

1. Constrain the translation in several ways:

   a) fix one or more coordinates to a pre-defined or the current value,

   b) keep the distance to other objects (e.g., ground) to a pre-defined or the current value. The distance is evaluated based on a direction which can be specified.

   This can be used to fix the user to eye level, for terrain following, or to make the user ride another object (an elevator, for example).

2. Constrain the orientation to a certain axis and possibly the rotation angle to a certain range. This can be used to create doors and car hoods.

All constraints can be expressed either in world or in local coordinates. Also, all constraints can be imposed as an *interval* (a door can rotate about its hinge only within a certain interval). Interaction with those objects can be made more convenient if the deltas of the constrained variable(s) are restricted to only increasing or decreasing values (e.g., the car hood can only be opened but not closed).

Another constraint is the notion of *walls*, which is a list of objects that cannot be penetrated by certain other objects. This is very useful to constrain the viewpoint or to make some objects rigid and solid.

One of the most basic physical concepts is gravity, which increases „believability" of our worlds tremendously. It has been implemented in an object property „fall", which makes objects fall in a certain direction and bounce off „floor objects", which can be specified separately for each falling object.

Object selection. There must be two possibilities for specifying lists of objects: *hard-wired* and *user-selected*.

In entertainment applications, you probably want to specify by name the objects on which an action operates. The advantage here is that the process of interacting with the world is „single-pass". The downside is inflexibility, and the writing of the interaction script might be more cumbersome.

Alternatively, we can specify that an action should operate on the currently selected list of objects. This is more flexible, but the actual interaction with the world consists of two passes: first the user has to select some objects, then specify the operation.

User modules. From our experience, most applications will need some specialized features which will be unnecessary in other applications. In order to integrate these smoothly, our VR system offers „callback" actions. They can called right after the system is initialized, or once per frame (the „loop" function), or triggered by an event. The return code of these callbacks can be fed into other events, so user-provided actions can trigger other actions.

These user-provided modules are linked dynamically at run-time, which significantly reduces turn-around time.

It is understood that all functionality of the script as well as all data structures must also be accessible to such a module via a simple, yet complete API.

1.5.5 Examples

The following example shows how the point-and-fly navigation mode can be specified.

```
cart pointfly dir fastrak 1 \
    speed joint thumbouter \
    trigger gesture pointfly
cartrev gesture pointflyback
cart speed range 0 0.8
glove fastrak 1
```

The hood of a car can be modeled by the following lines. This hood can be opened by just pushing it with the index finger.

```
constraint rot Hood neg \
    track IndexFinger3 \
    axis a b to c d \
    low -45 high 0 \
    on when active collision Finger13 Hood
```

The following is an example of a library „function" to make clocks. (This assumes that the hands of the clock turn in the local xz-plane.)

```
define CLOCK( LHAND, BHAND )
timer LHAND cycle 60
timer speed LHAND 1
/* rotate little hand every minute by 6 degrees in local space
*/
objattr LHAND rot add local 6 (0 1 0) time LHAND 60
/* rotate big hand every minute by 0.5 degrees in local space
*/
objattr BHAND rot add local 0.5 (0 1 0) time LHAND 60
/* define start/stop actions */
Stop##LHAND : timer speed LHAND 0
Start##LHAND : timer speed LHAND 1
```

The ## is a concatenation feature of **acpp**. By applying the definition **CLOCK** to a suitable object, we make it behave as a clock. Also, we can start or stop that clock by the actions

```
CLOCK( LittleHand, BigHand )
action „StartLittleHand" when activated  speech „clock on"
action „StopLittleHand" when activated  speech „clock off"
```

1.5.6 Architecture

So far, we have considered only a few of many modules a complete VR system comprises. For sake of completeness, we will briefly give an overview of the architecture of a VR system [3].

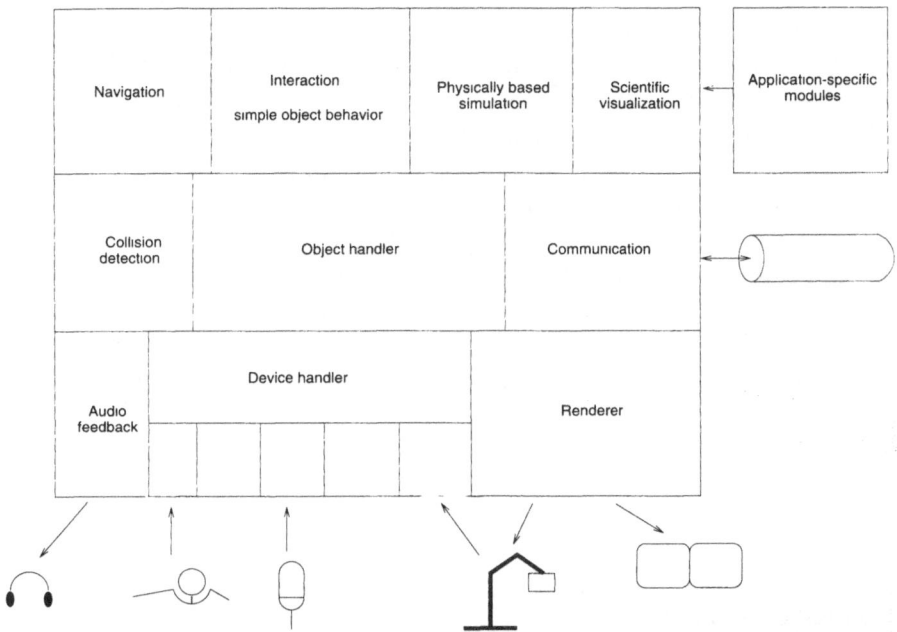

Figure 1.13. Architecture of a VR system. At the center is the object handler which maintains the scene graph. All modules must be able to run concurrently and asynchronously to each other, in order to achieve a constantly high frame rate.

Figure 1.13 shows a hierarchy of modules: at the bottom there are service modules which are primarily concerned with I/O, such as rendering, polling of input devices, and audio feedback. In the middle level there are service modules implementing functionality which is vital to many higher-level modules, such as collision detection and communication with other VR systems. The object handler is a supremely important module, because it maintains the scene graph. While most high-level modules keep their own module-specific, logical representations of objects, the scene graph is still *the* common basis to all of them. At the top level reside modules implementing functionality for interaction and simulation, such as navigation, object manipulation, physical behavior, and others.

The device handler implements a device abstraction by the notion of logical devices [9], a concept well known from **GKS** or **PHIGS**. It has proven quite practical to implement each device server so that it can be run on a remote machine and communicate with the device handler via a socket or other inter-process mechanism.

As stated above, virtually any non-trivial application needs some specialized functionality which will probably not be needed by any other application. This is depicted in the figure by the „application specific module", which is not a statically linked part of the VR system, but instead loaded by the VR system at run-time on demand. These modules still have full access to all data structures of the VR system.

A VR system must be able to sustain constantly a high frame rate (20 frames/sec or higher) under all circumstances, in order to achieve immersion and efficiency. Even more important is low *lag*, i.e., the latency between a change of the input data and a change in the output images must be minimal at all times. Therefore, modules must not block or delay the flow of data from the input devices to the rendered images. In order to achieve that, most modules depicted in Figure 1.13 run concurrently and asynchronously.

References

1. Akka R., Automatic software control of display parameters for stereoscopic graphics images, Unpublished results, 1992.
2. Andersson M., Carlsson C., Hagsand O., and Stahl O., DIVE – The Distributed Interactive Virtual Environment, Swedish Institute of Computer Science, 164 28 Kista, Sweden, 1994.
3. Astheimer P., Dai F., Felger W., Göbel M., Haase H., Müller S., and Ziegler R., Virtual Design II – an advanced VR system for industrial applications, In Proc. Virtual Reality World '95, pp. 337-363, Feb. 1995.
4. Blumberg B.M. and Galyean T.A., Multi-level direction of autonomous creatures for real-time virtual environments, In R.Cook, editor, Siggraph 1995 Conference Proc., pp. 47-54, Aug. 1995.
5. Clifford G.M.H., Shaffer A., A real-time robot arm collision avoidance system, IEEE Transactions on Robotics and Automation, 8(2), April 1992.
6. Cruz-Neira C., Sandin D.J., and DeFanti T.A., Surround-screen projection-based virtual reality: The design and implementation of the CAVE, In J.T. Kajiya, editor, Computer Graphics (SIGGRAPH '93 Proceedings), volume 27, pp. 135-142, Aug. 1993.
7. Dai F., Felger W., Frühauf T., Göbel M., Reiners D., and Zachmann G., Virtual prototyping examples for automotive industries, In Proc. Virtual Reality World, Stuttgart, Feb. 1996.

8. Feiner S., MacIntyre B., and Seligmann D., Knowledge-based augmented reality, Communications of the ACM, 36(7):53-62, July 1993.

9. Felger W., Fröhlich T., and Göbel M., Techniken zur Navigation durch Virtuelle Welten, In Virtual Reality '93, Anwendungen und Trends, pp. 209-222, Fraunhofer-IPA, -IAO, Springer, Feb. 1993.

10. Ghee S., dVS − a distributed VR systems infrastructure, In A.Lastra and H.Fuchs, editors, Course Notes: Programming Virtual Worlds, SIGGRAPH '95, pp. 6-1 - 6-30, 1995.

11. Ghee S., Mine M., Pausch R., and Pimentel K., Course Notes: Programming Virtual Worlds, SIGGRAPH '95, 1995.

12. Gottschalk S., Lin M., and Manocha D., OBB-Tree: A hierarchical structure for rapid interference detection, In H.Rushmeier, editor, SIGGRAPH 96 Conference Proceedings, Annual Conference Series, pp. 171-180. ACM SIGGRAPH, Addison Wesley, Aug. 1996, held in New Orleans, Louisiana, 04-09 August 1996.

13. Halliday S., and M.Green, A geometric modeling and animation system for virtual reality. In G.Singh, S.Feiner, and D.Thalmann, editors, Virtual Reality Software and Technology (VRST 94), pp. 71-84, Aug. 1994.

14. Hodges L.F., and Davis E.T., Geometric considerations for stereoscopic virtual environments, Presence, 2(1):34-43, winter 1993.

15. Jacoby R.H., and Ellis S.R., Using virtual menus in a virtual environment, ACM Computer Graphics, Siggraph '92, Course Notes(9), 1992.

16. Kirk D., editor, Graphics Gems III. Academic Press, Inc., San Diego, CA, 1992.

17. Liang J., Smoothing and prediction of isotrak data for virtual reality systems, In Proceedings of the 1991 Western Computer Graphics Symposium, pp. 58-60, Apr. 1991.

18. Lipton L., The CrystalEyes Handbook, StereoGraphics Corp.

19. Mackinlay J.D., Card S.K., and Robertson G.G., Rapid controlled movement through a virtual 3D workspace, In F.Baskett, editor, Computer Graphics (SIGGRAPH '90 Proceedings), volume24, pp. 171-176, Aug. 1990.

20. Maiocchi R., and Pernici B., Directing an animated scene with autonomous actors, The Visual Computer, 6(6):359-371, Dec. 1990.

21. Nam Y., and Wohn K.Y., Recognition of space-time hand-gestures using hidden Markov models, In Proc. of the ACM Symposium on Virtual Reality Software and Technology (VRST '94), Hong Kong, July1-4 1996.

22. Oishi T., and Yachi S.,Methods to calibrate projection transformation parameters for see-through head-mounted displays, Presence, pp. 122-135, winter 1996.

23. Pausch R., Burnette T., Brockway D., and Weiblen M.E., Navigation and locomotion in virtual worlds via flight into Hand-Held miniatures, In R.Cook, editor, SIGGRAPH 95 Conference Proceedings, Annual Conference Series, pp. 399-400, ACM SIGGRAPH, Addison Wesley, Aug. 1995, held in Los Angeles, California, 06-11 August 1995.

24. Riedel O., and Herrmann G., Virusi: Virtual user interface − Iconorientierte Benutzerschnittstelle für VR-Applikationen, In Virtual Reality '93, Anwendungen und Trends, pp. 227ff. Fraunhofer-IPA, -IAO, Springer, Feb. 1993.

25. Robinett W., and Holloway R., Implementation of flying, scaling, and grabbing in virtual worlds, In D.Zeltzer, editor, Computer Graphics (1992 Symposium on Interactive 3D Graphics), volume25, pp. 189-192, Mar. 1992.

26. Slatre M., Usoh M., and Chrysanthou Y., Virtual Environments '95, Selected papers of the Eurographics Workshop in Barcelona, Spain 1993, and Monte Carlo, Monaco 1995, pp. 8-21, Springer, Wien, New York.

27. Southard D.A., Transformations for stereoscopic visual simulation, Computers & Graphics, 16(4):401-410, 1992.
28. Taylor R.M., II, Robinett W., Chi V.L., Brooks F.P., Jr., Wright W.V., Williams R.S., and Snyder E.J., The Nanomanipulator: A virtual reality interface for a scanning tunnelling microscope, In J.T. Kajiya, editor, Computer Graphics (SIGGRAPH '93 Proceedings), volume27, pp. 127-134, Aug. 1993.
29. Turk G., Interactive collision detection for molecular graphics, Master's thesis, University of North Carolina at Chapel Hill, 1989.
30. Väänänen K., and Böhm K., Gesture driven interaction as a human factor in virtual environments, In Proc. Virtual Reality Systems, University of London, May 1992.
31. Wang Q., Green M., Shaw C., EM – an environment manager for building networked virtual environments, In Proc. IEEE Virtual Reality Annual International Symposium, 1995.
32. Ware C., and Osborne S., Exploration and virtual camera control in virtual three dimensional environments, In R.Riesenfeld and C.Sequin, editors, Computer Graphics (1990 Symposium on Interactive 3D Graphics), volume24, pp. 175-183, Mar. 1990.
33. Zachmann G., A language for describing behavior of and interaction with virtual worlds, In Proc. ACM Conf. VRST '96, July 1996.
34. Zachmann G., Distortion correction of magnetic fields for position tracking, In Proc. Computer Graphics International (CGI '97), Hasselt/Diepenbeek, Belgium, June 1997. IEEE Computer Society Press.

2 Immersive User Interaction Within Industrial Virtual Environments

Matthias Buck
Daimler-Benz AG

In this contribution we address immersive virtual reality (VR) applications, where the user enters the virtual environment (VE) entirely, and interacts with its elements as if being a part of the virtual world. This stands in contrast to applications like typical flight simulators, where the user mainly interacts with the real elements of the simulator cockpit.

Immersive VR simulations are particularly useful in industrial applications ifergonomical questions are to be answered, or if the interaction of a human operator with a technical environment has to be studied, as will be shown by several examples.

In the second part, we will discuss the basic elements and tasks which are required for realistic interaction within virtual environments. This will show the necessity to simulate in the virtual environment a certain degree of physical object behaviour, and in particular collisions and mechanical contacts between virtual rigid objects.

In the third part, we will present elements and methods to obtain such realistic behaviour of virtual environments, and discuss their possibilities and restrictions.

2.1 Introduction

In many industrial applications, we are interested in the interaction between humans and the product in all the phases of the product life cycle such as design, production, sales, operation, and maintenance. These phases however are not independent from each other. The ease of operation and maintainability of a product for example have to be taken into account already at early stages of design and construction, where no physical models of the final product are available. It is therefore helpful to be able to simulate situations like assembly, operation, or maintenance as realistically as necessary. As far as human interaction is involved, virtual reality techniques are very well suited to realize such simulations. This will be illustrated by an overview of typical industrial applications in section 2.2.

To simulate human interaction with technical environments, we have principally the choice between two possibilities. The first one is to replace the human being by a simulated man model, also called a 'manikin'. Such manikins can easily be adapted to different body sizes, and allow to generate normalized test situations.

The second possibility is to immerse a real person into a virtual environment, which interacts directly with the elements of a simulated virtual world. This allows the test person to experience the simulated situations directly, and to judge immediately the feasibility and ergonomy of the interactive tasks at hand. With this approach, human factors like perception, fatigue, or skills are directly implied, and need not be simulated as would be necessary with the manikin approach. In our contribution, we will address this second case in more detail.

For immersive interaction within virtual environments, the real test person has to be provided with all the necessary interfaces to perceive the virtual environment appropriately, to move around in it, and to manipulate the virtual objects involved in the scene as if they were real. For a realistic simulation, we expect virtual objects to behave in many ways like real objects, which is not self-evident. The simulation of virtual worlds requires more than just realistic visualization - it requires also realistic physical behaviour of the virtual objects. Realistic here not necessarily means physically correct behaviour in the strict sense, but at least physically plausible behaviour.

In the context of object manipulation, this means in particular that solid objects do not interpenetrate, that there are effects like friction, elastic shocks, and gravity, and that objects have appropriate mass and inertia properties. These effects require much more than just geometric representations of the virtual worlds. They require an appropriate physical simulation, and this of course with real-time response.

A specific difficulty of immersive interaction within virtual worlds is related to the feedback from the virtual environment to the user's physical world. When we work in real environments, we do not only see what happens, but we also feel the objects we touch. How can we feel virtual objects and the forces they excert on our hands? This indeed is a problem, which has not found a satisfactory solution so far.

The lacking of force feedback is not only a minor deficiency. As we will see later, reaction forces play an important role for the control of interactively manipulated objects, in particular if an object's motion is constrained by obstacle contacts. In the absence of satisfactory force feedback mechanisms, it is important to assist the user interaction in virtual worlds in a way that leads to an intuitively correct object behaviour nevertheless, and allows to work in a natural way, and with predictable results.

For readers, who are not familiar with the addressed subject, we give in section 2.2 an overview of typical industrial applications of immersive virtual environments to motivate the more technical sections that follow.

In section 2.3, we will discuss in detail the basic mechanisms which are required for immersive interaction within virtual environments, such as navigation, control, and feedback. In section 2.4 finally, we address methods for the simulation of a physically plausible object behaviour in virtual worlds.

2.2 Industrial applications

In the course of industrial product development, most decisions about the design, the user interface, the functionality, the mechanical construction, the planning of production and assembly lines, and the maintenance of the product have to be taken at early stages of the development cycle. Traditionally, these decisions are made on the theoretical basis of construction tools, CAD-systems, and hardware models at different scales, together with experiences from earlier development cycles. Many problems that emerge in the assembly, the operation or the maintenance of a product are recognized at rather late stages, in practical tests and during the application of the final product. Maintenance problems may even appear not before already many units have been sold. The fixing of such problems requires to loop back to the design and construction stage in order to do the necessary modifications. This procedure of course is inefficient, time consuming, and expensive.

The reason for such problems is that at early design stages, no practical experiences can be made with the product, because no prototypes are available for testing. The technical functionality in itself can thoroughly be tested and verified by simulations. It is however much more difficult to simulate and test the interaction between the technical product and a human operator in advance. It is just very difficult to decide only from CAD data whether a mechanician will have difficulties to assemble the part under construction, or to predict which difficulties a client may have in using the product.

Simulations in virtual environments offer a new chance to alleviate this problem. Even at early design stages, when the first computer models are already available, these models can be transferred into a virtual environment, and can be tested. As not only the appearance and the shape of the product can be simulated, but also the functionality behind it, chances are good that flaws and problems with the current design can be noticed at that early stage, and modified accordingly.

The following paragraphs illustrate the potential usage of immersive user interaction within virtual environments in certain industrial applications.

2.2.1 Ergonomy studies

The ergonomy of a product plays an important role for the customer satisfaction. A product, which can only be used with difficulty, or which requires the user to assume uncomfortable attitudes is less attractive compared to one with an ergonomically better design. If the user moves inside the product (e.g. a car or an aircraft), the perceived overall impression (tight or spacious cabin, good view outside) is an important criterion as well.

Figure 2.1. In an immersive virtual environment, like a simulated car interior, the user can experience if the instruments can be seen clearly, and if all control devices can be reached comfortably.

Ergonomical studies in virtual prototypes allow in early design stages to verify whether the „user interface" of a product (imagine an aircraft cockpit, the driver's seat of a car, a control panel, a work cell in an assembly line or in maintenance) is user friendly or not. The typical operations and procedures can be carried out in a way which is close to reality. The user has to make all the necessary movements, reach the control buttons, read the instruments, etc (see). Ergonomical flaws are very probable to be detected in an immersive virtual environment, and can be corrected (and be tested on the fly again) at an early design stage, which saves a lot of effort.

2.2.2 Assembly simulation, maintenance simulation

In an immersive interactive assembly simulation, two ergonomical aspects can be studied.First, the mechanician should be able to do the assembly in a way which is not uncomfortable, tiring, or even dangerous. Second, in certain products many parts have to be assembled in a small space, and it can be difficult or even impossible to mount some of the parts because other parts are in the way. Even if a part can be placed correctly, it may be difficult to put the screws and fix them. The same type of problems can occur during maintenance, where one particular part has to be replaced. It can be either difficult to reach the defective part, or it may even be necessary to remove other parts first before the broken part can be accessed. In all these cases, many potential problems can be detected early by simulation in immersive virtual mock-ups.

2.2.3 Design

The traditional design of a product like a car requires a lot of effort. Many wooden or clay models at different scales are required until the final shape has been determined. Such models are expensive as well as time consuming, and they are not flexible at all. This makes the comparison of alternative solutions complicated and inefficient.

Simulations with virtual environments are a very efficient means to accelerate the design process. Today, the virtual design models are inspected in immersive virtual environments, and design variations can be compared on the fly with realistic colours and illumination effects. Reflection effects can be simulated in real-time today (see Figure 2.3), which is an important tool to judge surface details and to detect even minor shape defects. One could imagine in the future in addition to modify object shapes interactively within the virtual world, like reshaping a clay model.

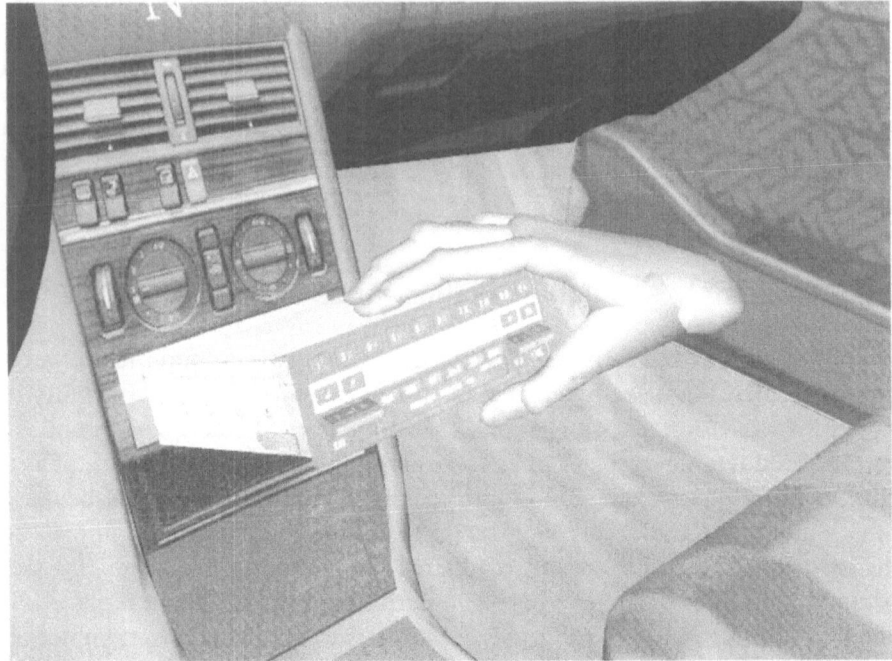

Figure 2.2. Ergonomic study in a maintenance simulation: The car radio is inserted interactively into the console using a data glove. Collision detection and contact simulation methods (see section 2.4 and Color Plates) assist the user at this task.

2.2.4 Digital mockup support

The 3-dimensional arrangement of modules in complex mock-ups can be very confusing and difficult to understand or to modify. Immersive virtual reality techniques allow an engineer to dive into the arrangement at any scale, and judge whether the packaging is satisfactory or not. If it is not, techniques for direct interactive manipulation within the virtual mockup can be used to fix the problem.

Figure 2.3. The quality of the design can be inspected interactively by the simulation of reflections of linear light sources (see section Color Plates).

2.2.5 Tele-operation

In certain applications, e.g. telerobotics in space, a device is not directly accessible and can only be controlled from a remote place. If the transmission channel for control and feedback data implies a time delay of more than a few hundred milliseconds, a direct interactive control is not possible. An example is the remote control of scientific experiments in an unmanned space station. Even in an orbital mission, time delays can accumulate to several seconds. For such applications, the real environment in orbit can be reproduced by a virtual copy on ground, in which the operator is working.This virtual environment reacts without delay to the actions of the operator. These actions and commands are transmitted (with delay) to the space station and executed there. A prerequisite of tele-operation is a good correspondence between the local virtual environment and the remote real environment. Any deviations between the two have to be corrected, which can be done by sending a video signal from the remote location to the control site, which allows to compare the real and the virtual world, and to update the virtual model if necessary.

2.2.6 Training applications

Training in real environments can be dangerous or very expensive (think of surgeons, aircraft pilots, astronauts). Wrong decisions or missing skills can lead to catastrophic results. Some setups like production plants do exist only once, and are not available for training sessions. These disadvantages can be overcome by simulated training environments. Training simulation has already been used for a long time in some areas (e.g. pilot training), and is now considered for many more applications. It is not only much cheaper and much safer than training under real conditions, it can also be more efficient. Critical situations, which need more attention than standard ones, can be repeated at will. Training scenarios can be adapted to the individual demands of each trainee. Some results indicate, that VR training even yields better results than the real counterpart does. Finally, VR training offers entirely new chances, because phenomena can be visualized and dependencies can be experienced by the learner in a way which is not possible in reality, but which is well suited to acquire new knowledge.

2.3 Elements of immersive interaction within virtual environments

The application examples from the last section showed the potential use of immersive interaction simulation for industrial tasks. To realize such immersive simulations, certain functional elements have to be provided, which are discussed in this section.

These elements are required to allow the user to navigate within the virtual environment, to control components of the virtual environment (e.g. grasp virtual objects and to move them around), and to get sufficient sensory information from the environment to be able to interact with it in an intuitive and reliable manner. Not all of the elements which are present in physical environments are represented adequately in today's practical virtual environments. This is partly so because the complexity of some applications is too high for the computational power of available systems to obtain realistic simulations in real time. This is in particular true where physical behaviour of virtual objects is required. In section 2.4, methods for the simulation of some essential properties of physical objects are described. Another important issue in immersive interaction is the missing of force feedback, as we will see at the end of this section.

2.3.1 Navigation

To get the impression of being immersed into a virtual environment, it is necessary to be able to navigate within this world and to look at it from different perspectives. This requires first a way to specify an ego-motion of the human observer, and second to visualize the world correctly from each viewing position.

Navigation control can be realized by tracking the user's head position and orientation (see the following section). This means, the real head motions are acquired and used to control the view point from which the virtual scene is observed. Another navigation mode is called *walk-through* or *fly-through*, which allows the user to move around in the virtual world without moving actually in the physical world. There are quite a number of ways how motion direction and velocities of such fly-throughs can be specified. For instance, the motion direction could be identified with the tracked gaze direction, or with the orientation of a pointer device which is controlled by a data glove. A good overview on such techniques is given in [9].

Once a new view point has been specified, the rendering of the scene can be realized with standard computer graphics methods, e.g. those contained in graphic libraries like OpenGL [11]. Finally the rendered images have to be presented to the user appropriately. Again, there are several posibilities, like head-mounted displays (HMD), BOOM[TM1], or projection devices like a CAVE [6]. A comprehensive evaluation of different visualization devices can be found in [13].

---

[1] A BOOM (Binocular Omni-Orientation Monitor) is a mechanically suspended binocular monitor, which in contrast to an HMD, is manually held before one's eyes.

2.3.2 Tracking

Tracking is not only used for navigation, but for the determination of the spatial position and orientation of any real object, with the intention of controlling the location of its counterpart in the virtual environment. Typically, tracking is applied to the user's head (in combination with an HMD or a CAVE), or the user's hand (frequently in combination with a data glove or a similar device).

Tracking systems determine the position and orientation of an object either mechanically, by the use of magnetic fields, by ultrasonic devices, or by optical sensors. The choice depends on an evaluation of factors like tracking precision, work space dimensions, hardware cost, or application specific criteria.

2.3.3 Control

Navigation and tracking enable interactive specification of ego-motion or object motion in VR environments. To interact with virtual worlds means in addition to take influence on virtual objects, for instance to grasp virtual tools and to work with them, or simply to operate switches and buttons on a virtual control panel.

In the physical world, interaction is generally realized through mechanical contact: objects are *touched*, tools are *grasped*, buttons are *pushed*. Mechanical contact is based on the exchange of contact forces, which means that both, the human operator, and some virtual object, should be able to excert forces on each other. This would require force feedback mechanisms from the virtual world to the real one, which are generally not available in VR systems in a satisfactory way (see next section on feedback).

For this reason, object selection and manipulation is frequently realized through specific tools in virtual environments. Mechanical contact can be replaced by collision detection mechanisms, which indicate for example that the virtual hand intersects an object to be grasped. Another possibility is to use a kind of virtual laser beam (see Figure 2.4) to point with at an object, or simply to select an object by looking in its direction (see [9] for more details). Control can be realized further by voice input. In this way it may however sometimes be difficult to express precisely yet concisely which action should occur.

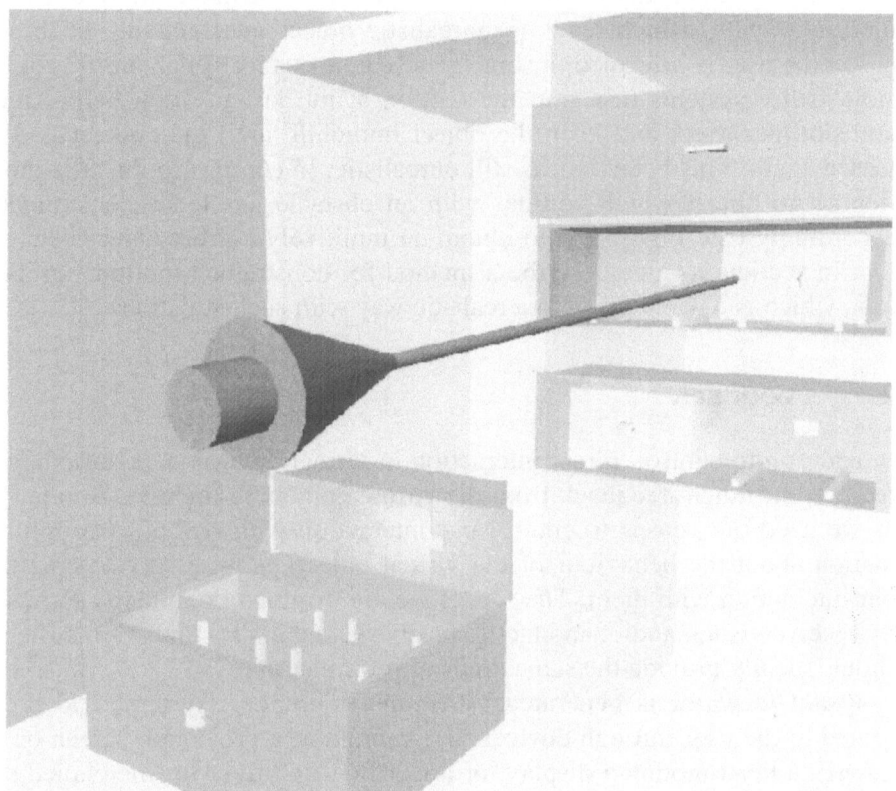

Figure 2.4. In virtual environments, even objects which are out of reach of one's arms can be selected and controlled by tools like a virtual laser beam.

By such means, actions like grasping an object and moving it around can be specified, even if the way this action is perceived by the user is quite different from the natural situation, as the feel of touch is missing.

The behaviour of controlled objects does however not only depend on the user's actions, but as well on constraints which are defined within the virtual environment. If the user specifies a certain motion for an object, which leads to a collision with other objects, the colliding geometries should not interpenetrate. Otherwise, the virtual environment would behave significantly different as the real one, and would not be suited for many simulation tasks. To prevent interpenetration, the originally specified object motion has to be modified automatically by the VR system in some way. Many VR systems provide mechanisms to detect such collisions, the subsequent consequences for the object's motion in general are however unrealistically simplified. In some cases, the object motion is not modified at all, which leads to unrealistic object intersections. In these cases, the user is informed in some way (e.g. acoustically) about the collision. Other systems just stop the motion at the last position before the collision occurred, and keep the object immobile until the contact is released again. This behaviour is still unrealistic. In contrast to this, the motion of an object which collides with an obstacle has to be constrained accordingly (see Figure 2.5) to obtain an intuitively correct object behaviour.In section 2.4, we describe a method for constrained motion simulation, which is a way to deal in a realistic way with such situations.

2.3.4 Feedback

In addition to control, direct interaction in virtual environments of course requires sufficient feedback from the virtual domain to the user. Similarly as we need our senses to control our interaction with real objects, information about the behaviour of the virtual objects is necessary to control our interaction with them. The senses we are using predominantly in reality receive visual, audio, and tactile feedback, and a VR simulation system should ideally provide the same kinds of information.

Visual feedback is generated by computer graphics systems and presented to the user through devices like a monitor, a projection screen or a CAVE, a head-mounted display, or a BOOM™. Criteria for the choice of the best suited device are image quality (resolution, contrast, frame rate), hardware costs, user acceptance, and application specific criteria. A good overview, and discussion of specific properties of all these devices can be found in [13].

From the point of view of direct manipulation in virtual environments, in addition to the mentioned criteria the perception of spatial depth is of particular importance. If we want to grasp an object in front of us, we require an estimation of where the object is located, and if we work in virtual environments we need 3D perception to avoid unwanted collisions with obstacles.This feature is supported by most VR systems by providing stereo views of the virtual scene. To each eye an individual image is presented. The two images are either displayed simultaneously on two screens (HMD, BOMM TM), or they are displayed on one screen using time multiplex or polarization techniques.

Audio feedback provides a rather general information of the scene, and one of its main benefits is that it increases the immersion effect of a virtual environment significantly.In the context of 3D navigation and interaction, spatial audio feedback can be a valuable help to orient oneself in a complex virtual environment. Furthermore, audio feedback can be used to notify the user about certain events, like object collisions.

Tactile or force feedback is one of the main senses humans rely on when manipulating objects. We can do difficile manual tasks (after some training) by only using tactile information, and even in the presence of visual feedback the tactile sense is still very useful. That is why already much effort has been made to construct force feedback mechanisms which provide realistic tactile perception, which are accepted by users, and which are meeting the requirements of industrial applications. We have to distinguish here between tactile feedback and force feedback. *Tactile feedback* means the sense of touch we feel with our skin sensors, in particular at our finger tips. *Force feedback* in contrast corresponds to what we feel in our arms if we e.g. are lifting heavy weights.Some modifications to data gloves have been developed to provide tactile feedback. They are based on subtle stimuli of the finger tips realized e.g. by small air cussions or vibration devices.On the other hand, all efforts to provide force feedback devices have not been very successful. Quite a number of experimental systems have been developed, which are either only suited for a restricted set of specific applications (e.g. for endoscopy training), or are not accepted by users (like exo-skeletons). In particular for assembly and maintenance simulations, many different parts have to be manipulated and moved across distances of several feet, and relatively large weight and contact forces have to be realized. For such requirements, no practical solutions are in sight. A rather complete and comprehensive overview on available haptic displays can be found in [14].

2.4 Realistic simulation of rigid objects with obstacle contacts

Experiences from existing interactive VR systems [7] show that if a VR application requires direct manipulation of virtual objects by the user, a certain amount of physically plausible behaviour of the virtual world is necessary to obtain satisfactory results. To illustrate this consider for example the way object collisions within the virtual world should be treated. Obviously, collisions between a manipulated virtual object and an obstacle should be recognized. Otherwise you could never carry out even simplest tasks like e.g. to put a work piece onto a table. Collision detection is first of all a problem of computational geometry, which answers the question whether two given objects do intersect or not. If they do, a collision has been detected. This does not imply much physics. However, what consequences follow from the fact that a collision has been recognized between two objects? In many VR systems the only consequence (if any[2]) is to reset the moving object to the state before the collision. This lets colliding objects keep sticking on each other, until the objects are separated again. Real solid objects would not interpenetrate either, but would allow colliding objects to roll and slide along each other at the mutual contact. This behaviour, in fact, requires physical modeling of the contact situation and the involved contact forces (and, if need be, friction forces). Physical simulation of mechanical systems is not new ([1], [5], [8], [10], [12]). The challenge here is to integrate such simulation techniques into VR systems in a way which provides the required real-time response, and which is compatible with the interactive components of such systems.In the following sections, we describe a method for interactive motion simulation in the presence of obstacle contacts.

2.4.1 Constrained motion simulation

A rigid virtual object can move with 6 degrees of freedom (DOF) if it is not subject to motion constraints. Such constraints can be caused by contacts to other objects in the scene. Consider the example of a box which is interactively moved around, and which the user wants to put flat on a table (see Figure 2.5).Before the box touches the table, it may translate and rotate freely. After one vertex of the box has hit the table surface, one degree of freedom is lost, however the box may still rotate freely around that contact point. It the box touches the table with an entire edge, a second DOF is lost. The box still may rotate around the touching edge, or slide along the table. Finally, if the box lies flat on the table surface, a third DOF freedom is lost, and only planar motion is left. The contact situation between both objects determines in which way the relative motion between the two is restricted.

---

[2] Some systems just inform the user about collisions by highlighting or acoustic feedback, but do not prevent intersections at all.

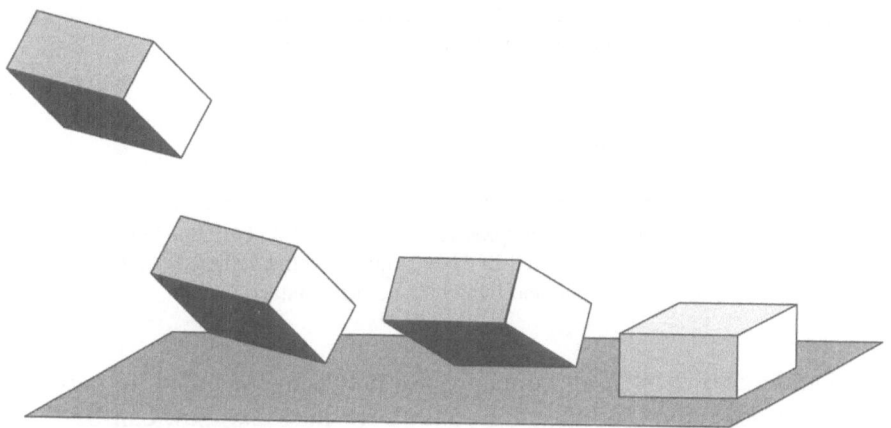

Figure 2.5. Obstacle contacts constrain the object's motion. For realistic interactive manipulations in virtual environments, this physical behaviour has to be simulated in the VR application.

This behaviour of real objects has to be simulated in virtual worlds if realistic object manipulation is required - which is the case in many of the industrial applications described above, e.g. in assembly simulations. Such simulation of contact situations and of the resulting motion constraints is a challenging task, and most VR systems do not contain this functionality, which requires to solve several difficult problems. It requires first a real time collision detection mechanism which determines all new contacts between any objects in the virtual scene. It requires second an ongoing analysis of the changing contact situations (see the box-and-table example), and the determination of a corresponding set of motion constraints for the involved objects. Finally, it requires a dynamics simulation scheme which determines the resulting constrained object motion of an interactively manipulated virtual object. All these functions have to be realized under real time conditions, such that the simulation reacts without noticeable delay to the user actions.

2.4.2 Contact analysis

Assuming polyhedral object geometries, there are a few basic contact types, which are combinations of the geometric elements vertex, edge, and face. The resulting 9 cases are shown in the table below, which indicates the number of DOF each contact type can remove from a pair of objects.

Contact region classification			
	face	edge	vertex
face	3	2	1
edge	2	1	1
vertex	1	1	1

Table 2.1. Classification of contact regions according to the number of degrees of freedom which they can remove from an object. In the example of Fig. 2.5, there is first a vertex-face contact, next an egde-face contact, and finally a face-face contact.

It can be shown, that all contact types between polyhedra, in particular edge-face contacts and face-face contacts (Figure 2.6 a,c), can be represented by a canonical (minimal) set of point contacts, i.e. by a set of vertex-face or edge-edge contacts (Figure 2.6 b,d). This set of point contacts can be determined and updated efficiently by evaluation of occurring collisions and breaking contacts.

2.4.3 Contact conditions and contact forces

At each contact point, a contact force enforces the non-penetration condition, i.e. it prevents the contacting objects from intersecting (see Figure 2.7).

The geometric contact conditions at contact i can be formulated as $d_i \geq 0$, where d_i is the contact distance. This distance of course is directly related to the moving object's location, given by the orientation matrix $\mathbf{R}$ and the translation vector $\mathbf{c}$, such that the contact condition leads to an implicit condition for the object's motion, $d_i(\mathbf{R},\mathbf{c}) \geq 0$. The object's motion in turn is related to the forces which act on the object through Newton's law, $f = \mathbf{ma} = m\dot{v}$.

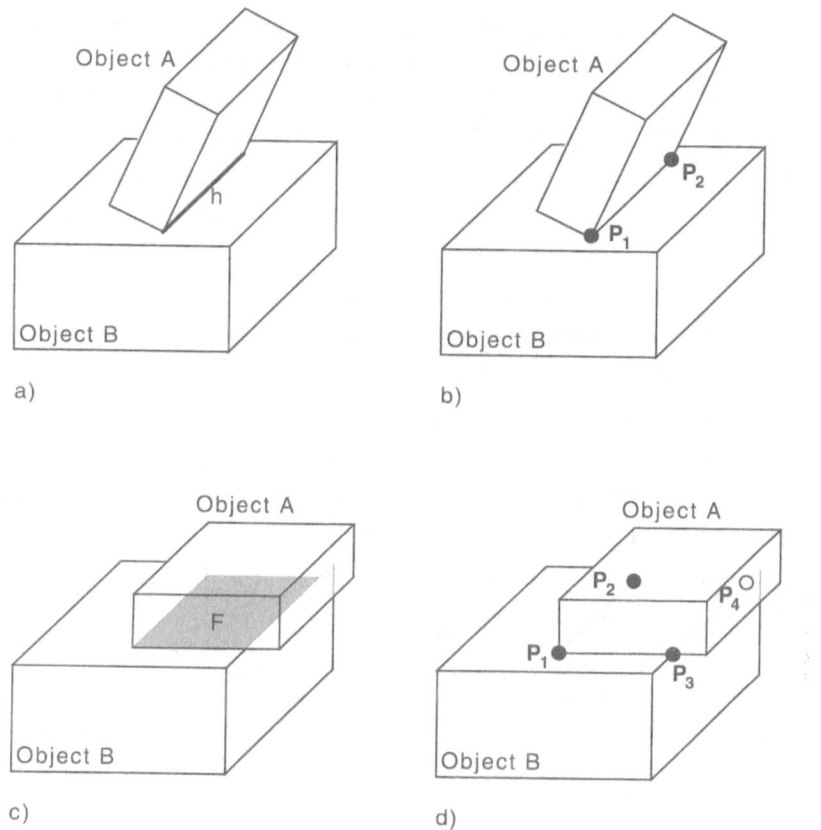

Figure 2.6. Distributed contact regions, like linear (a) or facial (c) contacts, can be approximated by few point contacts (b,d). P_1 and P_2 are vertex-face contacts, P_3 is an edge-edge contact. At these contact points, contact forces are simulated to prevent the intersection of the touching objects.

If several forces have to be taken into account, and if rotations are allowed, this leads to the Newton-Euler equations,

$$\dot{v} = \frac{1}{m}\left(F_{ext} + \sum_{i=1}^{k} n_i f_i \right)$$

$$\dot{\omega} = I^{-1}\left(M_{ext} + \sum_{i=1}^{k} r_i \times n_i f_i - \omega \times I\omega \right)$$

where v, ω are the translational and rotational velocities, m is the object's mass, I is the object's inertia tensor, and $n_i f_i$ are the orientation and magnitude of contact force i. F_{ext} and M_{ext} are the external force and torque applied to the object due to interactive control by the user.

After discretization of these differential equations and two numerical integration steps, we obtain from the accelerations $\dot{v}(f)$ and $\dot{\omega}(f)$ the orientation matrix $\mathbf{R}(f)$ and the translation vector $\mathbf{c}(f)$ as non-linear functions of the magnitudes of the contact forces, which have been collected in the vector $f = (f_1, f_2, ...f_n)$

To sum up, the contact conditions $d_i(\mathbf{R}(f), \mathbf{c}(f)) \geq 0$ can be reformulated as an implicit function of the contact forces, $d_i(f) \geq 0$. The contact forces themselves have to satisfy the conditions $f_i \geq 0$, as contact forces can only be repulsive. As the contact force f_i must be zero if the contact breaks (i.e. $d_i > 0$), and the contact distance d_i must be zero as long as a non-zero contact force $f_i > 0$ is acting, we obtain the so-called complementarity condition $f_i d_i = 0$.

This alltogether leads to a system of inequations for the unknown contact forces f,

$$d_i(f) \geq 0, \quad f_i > 0, \quad f_i d_i = 0, \quad i = 1, ...,n$$

where n is the number of point contacts. This mathematical problem is called a non-linear complementarity problem (NCP), which can be solved iteratively.

If the resulting contact forces are applied to the moving object, the contact constraints are satisfied, and the object motion corresponds to the expected physical behaviour.

This approach has been implemented at the Daimler-Benz Virtual Reality Competence Center (VRCC) by the author. The resulting computation times[3] are promising. For 4 contact points, the motion simulation took on average about 1ms. For 10 contact points (see Figure 2.8) the average was about 11ms. These results indicate, that the motion simulation with contacts can be used for real-time applications. It should be mentioned however, that the time needed for the collision detection (which is not included in above computation times) can be a computational bottleneck if the scene complexity is too high.

In the above derivation, only frictionless contacts have been assumed. To obtain a realistic object behaviour in virtual worlds, it may be necessary to include friction effects, as well as other effects like gravity and

---

[3] on an SGI-O2 with R10000, 150MHz

elastic shocks. Work on these topics is pursued at the VRCC as well, and results are expected in the near future.

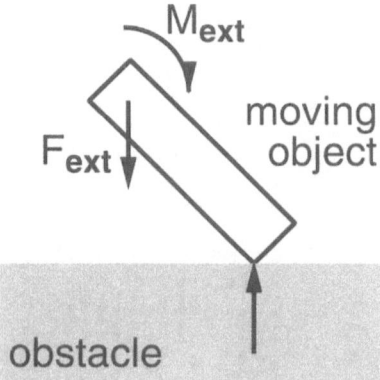

Figure 2.7. Contact forces prevent the moving object from diving into the obstacle surface. Together with the interactively controlled external forces they lead to the resulting object motion which respects the contact constraints.

To integrate all these components and demands into a reliable real-time framework makes up the challenge of realizing immersive interaction within virtual environments.

2.5 Conclusion

In our contribution, we discussed the importance of immersive user interaction within industrial virtual environments. A wide scale of industrial applications was presented to illustrate this, among them ergonomic studies, assembly simulation, and training.

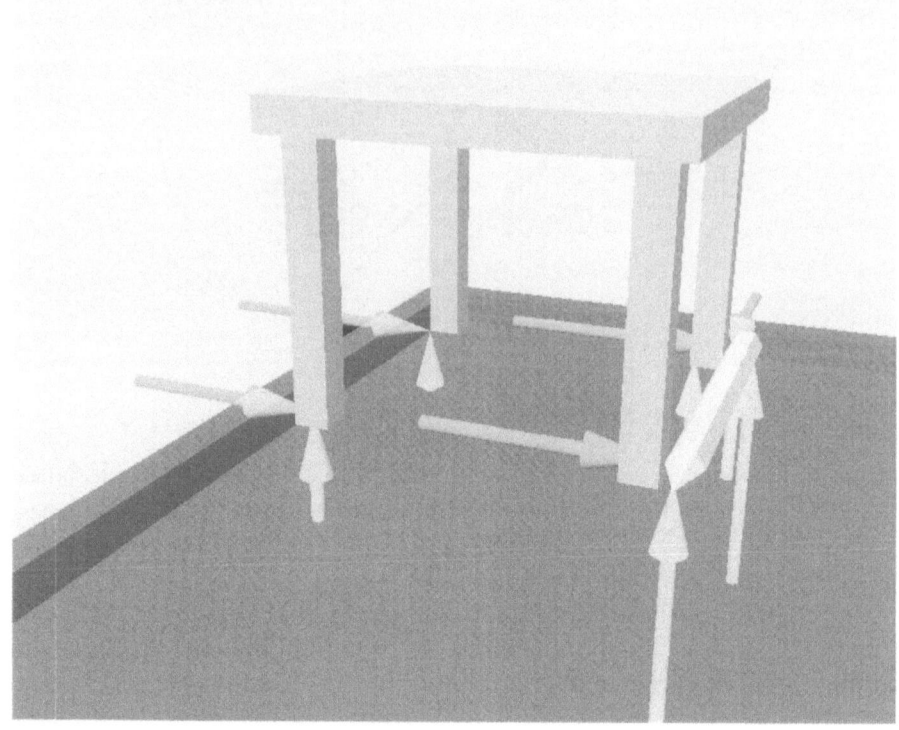

Figure 2.8. Visualization of the contact forces in a constrained motion simulation, showing two movable objects (a table and a rod) on a fixed support object. At each contact point, a contact force applies. The length of each arrow represents the magnitude of the contact force.

We further gave an overview of the methods and tools which are required for such immersive user interaction in virtual worlds, among them control and feedback mechanisms. An important drawback in practical systems, which has found no satisfactory solution so far, is the lacking of acceptable force feedback mechanisms. It was pointed out, that one important requirement for a meaningful use of such interactive simulations is an intuitively correct physical behaviour of the virtual world.

Most available VR systems do not provide even a basic physical behaviour, e.g. that solid objects can collide, and in consequence slide along each other without interpenetration. At the DaimlerBenz Virtual Reality Competence Center, methods are being developed to provide such physical behaviour. In particular, the problem of efficient collision detection and the simulation of motion constraints due to contacts between virtual objects are addressed. In the last part of this contribution, we presented a method for such constrained motion simulation in interactive virtual environments, which includes the simulation of sliding contacts between moving objects under real-time conditions.

References

[1] Baraff D., Dynamic simulation of non-penetrating rigid bodies, Ph.D. Thesis 92-1275, Cornell University, March 1992

[2] Baraff D. Linear-time dynamics using Lagrange multipliers, Technical Report CMU-RI-TR-95-44, Carnegie Mellon University, January 1996

[3] Bouma W., Vanêçek G., Contact Analysis in a Physically Based Simulation, ACM Symposium on solid Modeling and Applications, Montreal, Canada, May 1993

[4] Buck M., Grebner K., Katterfeldt H., Modeling and Interaction Tools for Virtual Environments, Proceedings of the VR World 96, Stuttgart, Germany, 1996

[5] Cremer J.F., Stewart A.J., The Architecture of Newton, a general-purpose dynamic simulator , IEEE International Conference of Robotics and Automation, pp.1806-1811, 1989

[6] Cruz-Neira C., DeFanti T.A., Sandin D.J., Surround-Screen Projection-Based Virtual Reality: The Design and Implementation of the CAVE Virtual Environment , Proceedings of SIGGRAPH '93, pp.127-135, July 1993

[7] Dai F., Felger W., Frühauf T., Göbel M., Reiners D., Zachmann G., Virtual Prototyping - Examples for Automotive Industry, Proceedings of the VR World 96, Stuttgart, Germany, 1996

[8] Lötstedt P., Mechanical systems of rigid bodies subject to unilateral constraints., SIAM Journal of Applied Mathematics, 42(2):281-296, 1982

[9] Mine M.R., Virtual Environment Interaction Techniques, Technical Report, Univ. of North Carolina, 1995

[10] Mirtich B., Canny J., Impulse-based dynamic simulation In Goldberg K., Halperin D., Latombe J.C., and Wilson R. (edts.), The algorithmic foundations of robotics, Peters A.K., Boston, MA, 1995

[11] Silicon Graphics Inc., The OpenGL Programming Guide - The Official Guide to Learning OpenGL, Addison-Wesley, Reading, Mass., 1993

[12] Vanecek G., Jr., Cremer J.F., Project Isaac: Building Simulations for Virtual Environments, Technical Report, Purdue University, June 1994

[13] Wloka D.W., CAVE: Personal or Small Group Non-HMD-based Head-Tracked Wrap-Around Virtual Environment - The System oh the Future for Virtual Reality Applications? , Proceedings of the VR World 96, Stuttgart, Germany, 1996

[14] Ziegler R., Haptic Displays - How can we feel Virtual Environments? , Proceedings of the VR World 96, Stuttgart, Germany, 1996

3 A Vision Based System Controlling the Virtual Tables Camera

Frank Seibert, Heike Kühner
Computer Graphics Center (ZGDV), Germany

The paper focuses on an adaptation of a virtual camera to a horizontal stereoscopic projection (Virtual Table). To get the correct view to stereoscopic projections it is necessary to detect the users position while walking around the table. One disadvantage of solutions based on magnetic trackers is their sensibility to distortions caused by external magnetic fields. Another unacceptable point is the obstructive cable connection of the user.

An alternative video based method has been developed which frees the user from carrying any special equipment (except of the stereo glasses). Its main requirements are robustness at real time conditions. The presented algorithm is based on the calculation of conditional probabilities by using chrominance histograms and additional methods to stabilise the detection results against noisy video signals and multiple observers included in a video image. It results in a combined x-, y-, and z-position enabling the rendering system to generate an adapted view of the stereoscopic scene matching the users position and the virtual camera of the table.

3.1 Introduction and objectives

3D stereoscopic displays become more and more common for the use in interactive environments. Besides traditional monitors, that limit the number of users because of its size, large-screen projections, CAVEs [Cru93], and horizontal output devices [Kru94, Fak96] are developed. By mapping high-volume, multidimensional data into meaningful displays and by enabling natural interaction with these displays, VR empowers users to get a realistic impression of the model, especially concerning its spatial connection.

The VIP project (VIP - Virtual Plane, ESPRIT 20642), which is supported by the European Commission, serves to develop a combination of horizontal display system and audio-visual interactive control. The Virtual

Table represents a horizontal output device that is build especially for applications conducted at tables, workbenches, or presentation platforms. Using shutter glasses, the users observe a 3D model rising above the tabletop. Figure 1 shows the projection of a three-dimensional object onto the Virtual Table.

Figure 3.1. Stereoscopic presentation on the Virtual Table (see section Color Plates).

In a stereoscopic display system, two perspectively shifted views of a scene are generated and displayed in such a manner that only the left eye sees the left eye's view and vice versa. The convergence of these two images yields the three-dimensional scene. Using these glasses does not imply a complete immersion of the user into the virtual world; he or she is still in contact to the real environment. As opposed to head-mounted-displays (HMDs), this helps to avoid motion sickness and disorientation. In addition, the visual perception of multiple users enables a natural communication between them.

One of the special qualities of the Virtual Table is that the horizontal projection surface provides a very natural way for presentation purposes and a basis for discussions. The tabletop metaphor allows a group of persons to look at the images simultaneously. The model can be evaluated by watching from different viewing directions and angles. Due to the unrestricted view of the other users, a very natural communication is possible.

Such devices provide a human-computer interface that is different from using a monitor or vertical-based device. This device is not necessarily better, just because of a changed human-computer interface. If the interface is used effectively, however, it improves the nature of working at desks, tables or workbenches. In contrast to vertical-based display technology the traditional *fly-through*-metaphor does not exist any longer for horizontal output devices.

Applications that fit into this environment also differ in regard to their interactions. The goal is to project a virtual model onto the projection tables surface, with the users walking around the table to watch the model from different angles. It is highly important to keep the image still during this procedure. The model stays in its original position or rather in the tables dimensions, which corresponds to the natural way of watching things on a table. This aspect is faced with tracking the users' motion and with rendering the correct perspective from the different viewing angles. This does not imply, however, that the object cannot be moved; it can certainly be translated, rotated, and scaled with the help of proper interaction metaphors.

Working with the horizontal display several aspects needed to be evaluated carefully in order to use this technology efficiently. Beside the adapted perspective the question on how to interact with the projected scenario needed to be evaluated.

Sufficient space for discussion is important for collaborative aspects existing using the Virtual Table. In this regard, wired input devices for interactions with the projected three-dimensional model have always been a strong limitation, since cables tend to get stuck at the borders and users must be careful not to get caught by cables when moving. The first step to overcome these limitations was the development of the video-based head tracking system DISCUS [Wol97].

Experiences with the Virtual Table have shown that the two aspects: adaptation of the projection matrix to the horizontal output surface and the necessity of wireless input devices for head tracking and intuitive interaction purposes are needed for the use of this kind of devices. In this paper first we describe the basic idea to calculate the correct perspective. The probability based segmentation technique combined with an object tracking and stabilising mechanism for the detection of the users position is introduced in more detail.

3.2 Visualisation techniques

An adaptation of the virtual camera was essential for ensuring the correct representation of the model from any viewpoint. In contrast to vertical output devices, the viewing angle is variable and corresponding to the users position (see figure 2). The dotted line shown in figure 2 represents the viewing angle having a vertical projection.

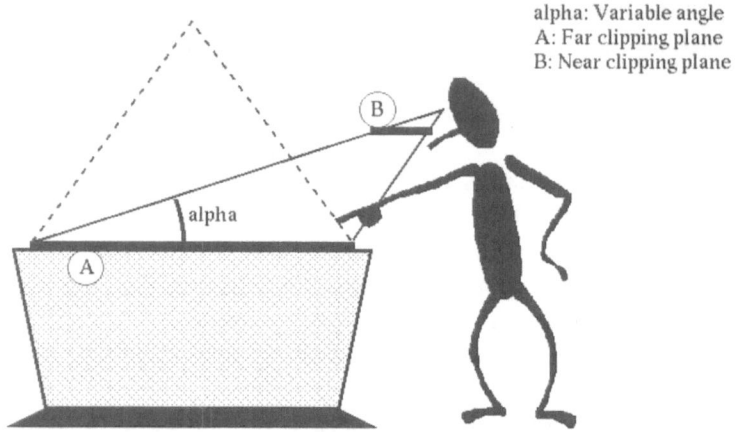

alpha: Variable angle
A: Far clipping plane
B: Near clipping plane

Figure 3.2. Side elevation of the adapted viewing volume

Therefore the viewing volume was changed in such a manner that it possibly adopts an asymmetric shape in dependency of the users position. The goal is to project a virtual model onto the projection tables surface with the users walking around the table to watch the model from different angles. It is highly important to keep the image still during this procedure. The users movement around the table is evaluated by a tracking mechanism (see figure 3). The images are calculated in such a manner that the bottom part of an object is placed exactly on the surface of the projection screen. This means that the far clipping plane is placed exactly on the surface of the projection table [Fol90]. Doing this we get the impression that everything is standing on the tables surface. Also the far clipping plane is not changed with the users movement. while the near clipping plane is changing its position in dependency of the users position. Both clipping planes far and near are marked with A and B. The distance between these planes is also dependent on the users position.

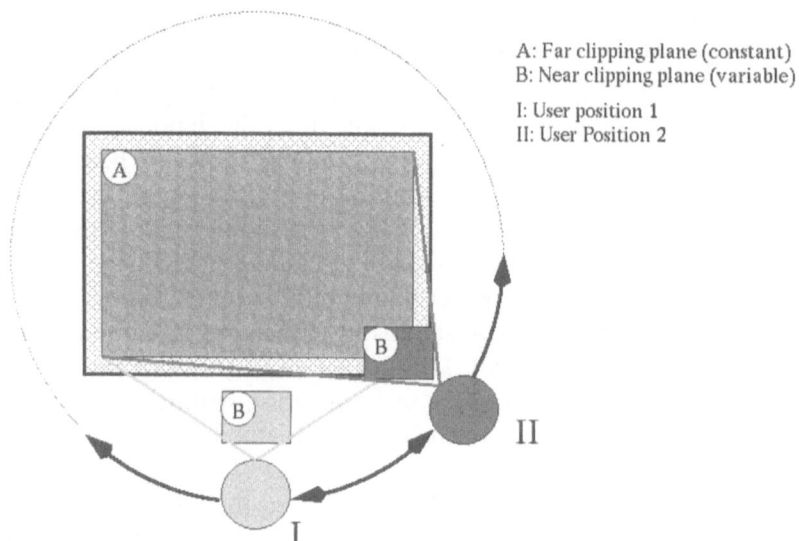

A: Far clipping plane (constant)
B: Near clipping plane (variable)

I: User position 1
II: User Position 2

Figure 3.3. Adapted viewing volume of different viewing angles for different user positions

3.3 Interaction

Human-computer interfaces play a role of growing importance as computer technology continues to evolve. Indeed, for many applications, the
productivity bottleneck lies with the user's ability to communicate with
the computer rather than with the CPU's limitations.

Traditionally, in VR applications magnetic or mechanical tracking systems are used for the evaluation of the users position [Bry96]. However,
these technologies suffer from various problems, including cost, lack of
accuracy, distortion sensitivity by the environment, and limited range.

Our experiences with a magnetic tracking system have shown that interaction with the objects displayed on the Virtual Table suffer from the
distortions caused by the external magnetic fields of the stereo projectors
and the tables aluminium frame. Beside of the distortions, to use a magnetic tracking system means that the user is wired. Especially when using
the Virtual Table wired input devices have always been a strong limitation, since cables tend to get stuck at the borders and users must be careful
not to get caught by cables when moving. To overcome these limitations
we adapted the camera based head tracking system DISCUS.

Initially, this tracking system was developed for the use at traditional monitors. Displaying three-dimensional objects onto a two-dimensional monitor leads to missing information. For example, watching an object from the front does not give any information about the depth of the object. A very natural way for getting this information is to move the head to look from a different position. Using a 2D mouse in a virtual environment this traditionally is simulated by the rendering process with rotating the object relative to the movement of the mouse. The camera based head tracking allows the user to discover the object in a natural way. In contrast to a 3D perception which is based on a stereoscopic projection onto a screen where special glasses are needed, DISCUS is used to adapt the viewing angle to a 3D object by detecting the users position without need for carrying additional equipment. Nevertheless this method may be combined with a stereoscopic projection to get a maximum of 3D impression.

Since first tests with the Virtual Table clarified the need of a wireless head tracking, this system was adapted to the requirements of projecting VR applications onto a horizontal output device.

3.3.1 Camera based head tracking

For the use of a camera based head tracking within a non-immersive environment the system needs to be robust against changing backgrounds, signal noise [Aac93], and variable lighting conditions. To allow multiple observers to work on the displayed model is one of the Virtual Tables special qualities. For that other people may also be located within the range of the camera and may walk around the projection table. This aspect has to be faced by the tracking system.

Individual characteristics like a beard or glasses should be irrelevant for the detection of the users position. At the same time not just one user should be able to perceive the correct perspective which means to be tracked. The handing over of the focus during a session is rather wanted.

To interact within a virtual environment users are often forced to learn about special mechanisms. Calibration may sometimes be very complicated and time consuming. Systems based on Computer Vision techniques sometimes need to attach markers to the users body. The necessity of markers attached to the users body may lead to non-acceptance of the system. The camera based head tracking described in the following sections is considering the mentioned requirements.

Different ways exist to detect the users position in a video sequence. We considered four auspicious methods (matching, motion detection, probability segmentation, and classification) and valued them in relation to fulfil the requirements of the given scenario.

The advantages regarding the Virtual Tables requirements in using a probability based segmentation technique are the following:

- the background can change,
- the lighting conditions can change (as long as the chrominance values stay almost constant),
- low computing effort.

Difficulties with the calibration of the skins colour need to be solved and additional algorithms need to be implemented to handle camera noises. A probability based segmentation technique combined with an object tracking and stabilising mechanism has been selected to be the best choice. In the following sections the evaluation of the users position with regard to the different requirements is described.

Face segmentation

The segmentation separates the users face from the background. In our system this is a challenging task. Almost constant characteristics of a humans face had to be found. Also these characteristics should possibly not occur in other parts of the signal. The colour of the humans skin consists of a particular hue and saturation and most parts of the human face consist of skin. As a consequence we can say that complexion is a suitable characteristic for separating the users face from the background.

In a first step we transform the video image into a face probability image. As already mentioned one requirement for the usability of the tracking system was its independence against lighting conditions. To face this requirement a colour model has to be selected which separates the luminance information from the colour information. Since the colour information will be computed, the YUV model which is represented by the UV-Vector fits. According to equation 1 the given RGB signal is transferred into the $C_B C_R$ space [Poy95] in order to get a colour vector where both components U and V have the same range.

$$\bar{C} = \begin{bmatrix} C_B \\ C_R \end{bmatrix} \approx \begin{bmatrix} 128 \\ 128 \end{bmatrix} + \frac{1}{255} \begin{bmatrix} -38 & -74 & 112 \\ 112 & -94 & -18 \end{bmatrix} \times \begin{bmatrix} R \\ G \\ B \end{bmatrix} \quad (1)$$

$\bar{C}$ represents a possible result with the conditional probabilities $P(\bar{C} | H_0)$ and $P(\bar{C} | H_1)$ of the following hypotheses:

H_0: Pixel belongs to the users skin

H_1: Pixel does not belong to the users skin $\qquad$ (2)

With the goal to transform the video image into a face probability image the probability $P(H_0|\bar{C})$ can be calculated according to the rule of Bayes [Cro95]:

$$P(H_0|\bar{C}) = \frac{P(\bar{C}|H_0) \times P(H_0)}{P(\bar{C}|H_0) \times P(H_0) + P(\bar{C}|H_1) \times P(H_1)} \quad (3)$$

The denominator represents the probability $P(\bar{C})$, because of the two alternative hypotheses. This leads to equation 4. $P(H_0|\bar{C})$ represents the conditional probability of chrominance vector $\bar{C}$ belonging to the users skin. Knowing this probability we are able to transform the video image into a face probability image.

$$P(H_0|\bar{C}) = \frac{P(\bar{C}|H_0) \times P(H_0)}{P(\bar{C})} \quad (4)$$

Dividing the number of pixels having this chrominance by the number of pixels of the complete image the probability $P(\bar{C})$ can be calculated with the current chrominance histogram. To calculate $P(\bar{C}|H_0)$ with the current chrominance histogram we need to know which pixel belongs to the users skin and which pixel doesn't belong to the skin. Presupposing that the chrominance values of the skin are approximately constant for one application it is sufficient to evaluate the conditional probability once within an initialisation. The probability $P(H_0)$ can be estimated by the calculated number of pixels belonging to the users skin within the previous video frame.

The initialisation of the head tracking system is carried out simply by forcing the user to cover a specified region within the video sequence in order to memorise the faces colour. The rectangle representing the calibration region is shown in Figure 3.4.

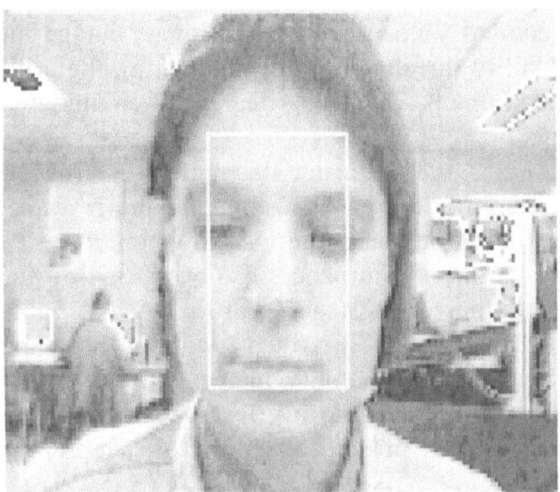

Figure 3.4. Calibration of the chrominance value of the users skin

It is $P(H_0) = 1$ for each pixel inside the initialisation region. This means that the probability $P(\bar{C}|H_0)$ can be calculated by dividing the value within the chrominance histogram by the number of pixels within the region. The chrominance histogram includes the number of pixels within the specified region together with the corresponding chrominance values for each possible colour vector.

In comparison to the use of the camera based head tracking at traditional monitors the size of the initialisation region needs to be adapted to the larger distance of the users to the video camera.

Evaluation of the face position

The first step was to transform the video image into a face probability image. The next step focuses on the evaluation of the users x-, y- and z-position. Due to the fact that the video image is two-dimensional it has to be considered that the z-coordinate is represented by the distance between camera and user. An attribute has to be evaluated which describes this distance. In our context we are computing the faces size.

For the evaluation of x- and y-position we defined to use the pixel position of the video image as the coordinate system. Furthermore we need to define which pixel belonging to the skin is representing the users position. The distribution of pixels belonging to the skin are approximately of the shape of an ellipse. Now we are saying that the centre of gravity of this ellipse represents the users x- and y-position. After a binary image has been produced using a probability threshold we are calculating the arithmetical mean of all face related pixels. This is carried out in combination with a tracking mechanism.

By calculating the centre of gravity only inside this region disturbances outside this region are filtered. In addition, by estimating the restricted moving possibility of the user a region can be defined in which the users face is still present in the next frame. Disturbances can be caused by surrounding noises as well as by other users standing at the Virtual Table. Figure 5 shows the conditional face probability image without the search region, figure 6 the conditional face probability image with the search region. The left image of the two figures are overlayed with the actual video image while the right one is pure presented. The crossbar represents the centre of gravity.

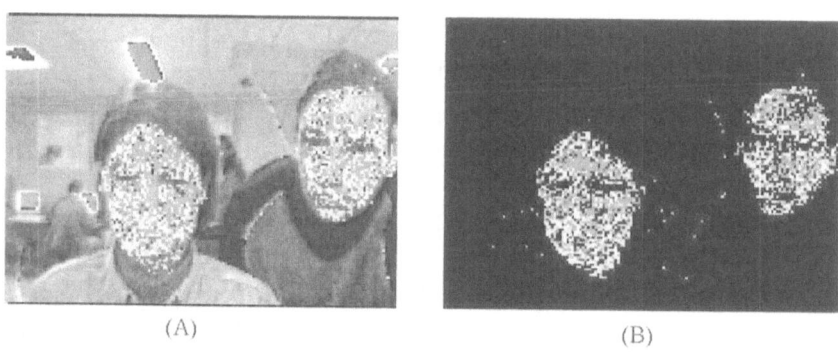

(A)   (B)

Figure 3.5. Face probability image without search region

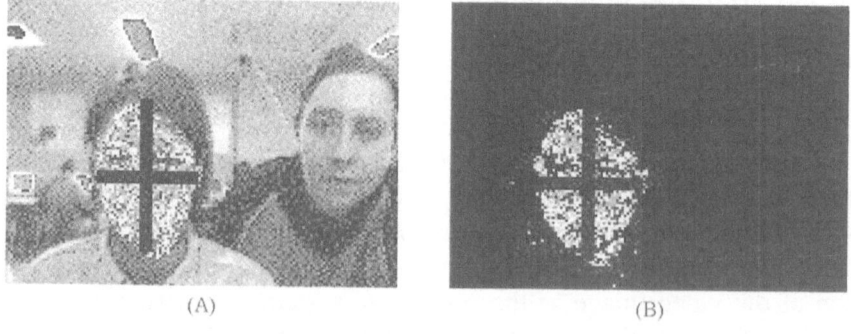

(A)   (B)

Figure 3.6. Face probability image with search region

The evaluation of the z-coordinate is represented by an approximation of the face size within the video image. In [Wol97] three methods are valued in regard to accuracy and performance. Two of those are implemented and are available optionally. In comparison to the density method the standard deviation method was considered to be the best choice, because the computational effort is much smaller. Using the density method the computational effort is higher, but the accuracy is better.

The calculated x-, y- and z-position of the user represents the relative position inside the current video image. Transmitting these positions to the VR system and rendering the model according to these values would deliver bouncing objects. The reason is the noise of the video signal. The stabilisation that will be described in the following section was implemented to avoid this situation. To use the camera based head tracking within interactive environments where real time is an important aspect this disadvantage absolutely needs to be solved, especially for situations where users don´t move while varying detection results will change the view.

Stabilisation

As already mentioned, problems occur as soon as the user keeps still in a certain position. What happens is that the same position has to be detected in each image of the video sequence during this time period. This is not realistic using a statistical based segmentation method where rest disturbances are certainly appearing.

To face this problem we implemented an arithmetical stabilisation that smoothes detected positions by calculating its arithmetical mean in a certain period. The result of stabilisation is better the higher the number of regarded frames is. But regarding a large number of frames causes latency of the rendered image if the user moves faster.

A dynamic stabilisation solves this problem. The degree of stabilisation is decreased as soon as the user moves faster and vice versa. The standard deviation of previous position or the actual velocity of the user can represent an indicator for the changing of stabilisation degree.

The stabilised relative values of x-, y- and z-position are transmitted to the VR system via a socket communication and transformed into the three-dimensional tables coordinate system. Knowing the users position the correct perspective for the Virtual Table is calculated and visualised.

3.4 Summary and future work

This paper described the first steps necessary for the effective work with a horizontal output device. Due to the table top metaphor of the Virtual Table traditional work procedures as well as collaborative work is supported. If this interface is used efficiently, however, it improves the nature of working at desks, tables or workbenches. Also the acceptance of this technology by the user is improved.

In order to build up applications that fit into this environment the necessary adaptations and implementations in the visualisation- and interaction mechanisms needed to be developed. These mechanisms face the features of the horizontal projection surface. We first introduced the basic idea of the visualisation technique in order to compensate perspective distortions caused by different user positions which makes head tracking necessary. In order to visualise perspectively correct images the virtual camera was adapted to the horizontal projection surface. This was necessary to generate the impression that the virtual objects are floating above the projection surface.

Furthermore, experiences with the head tracking at the Virtual Table clarified the need of a wireless tracking system. We described a camera based head tracking system which was integrated into the VR system. Using this kind of tracking system a good possibility for the use at the Virtual Table is provided. The wireless evaluation of the users position facilitates collaborative work at the projection surface where multiple observers may walk around and discuss the presented model. One drawback of the system is the missing detection of the users orientation. In combination with the Virtual Table this means that the view point is fixed to the center of the projection screen. In order to provide the detection of different focus areas on the projection plane the evaluation of the heads orientation needs to be realised. This could be done with the extraction of more characteristics of the users face beside of the skins colour. For example, this could be the knowledge about the relation between eyes and nose.

Humans are natural experts at manipulation in two- and three-dimensional spaces. Gesture inputs aim to exploit this natural expertise in order to bring the human-computer interface functionally closer to the human physical and human-human interfaces. The idea here is to make explicit use of mechanisms which humans have learned and experienced during their whole life in the manipulation of objects. To provide users with an intuitive gesture input we made first experiments with a camera based gesture recognition. The users gestures are recorded by a video camera and evaluated and classified by means of computer vision techniques. Two advantages of this technique in contrast to e.g. datagloves are the wireless control of input devices and the avoidance of a virtual representation of the users hand, which reflects the manner of working at a table or workbench quite naturally.

References

[Aac93] Aach T., Kaup A., Mester R., Statistical model-based change detection in moving video. Signal Processing 31, 1993.
[Bry96] Bryson S., Virtual Reality in Scientific Visualisation. Communications of the ACM, Vol. 39, No. 5, May 1996.
[Cro95] Crowley J. L., Coutaz J., Vision for Man Machine Interaction. ECVNet 1995.
[Cru93] Cruz-Neira C., Sandin D.J., DeFanti T.A., Surround-Screen Projection-Based Virtual Reality: The Design and Implementation of the CAVE. In Computer Graphics Proceedings, Anual Conference Series, 1993, pp. 135 - 142.
[Fak96] Fakespace/SGI Debut, The Immersive Workbench. Real Time Graphics, Vol. 5, No. 4, October/November 1996.
[Fol90] Foley J., van Dam A., Feiner S., Hughes J.: Computer Graphics: Principles and Practice. Addison-Wesley Publishing Company, Second Edition, 1990.
[Kru93] Krüger W., Bohn C.-A., Fröhlich B., Schüth H., Strauss W., Wesche G., The Responsive Workbench: A Virtual Work Environment. IEEE Computer Graphics and Applications, May 1994, Volume 14, Issue 3, pp. 12-15.
[Poy95] Poynton C.-A., Frequently Asked Questions about Color, Inforamp Net 1995.
[Wol97] Wolf M., 3D-Bildschirmausgabe durch kamerabasierte Benutzerinteraktion. Diplomarbeit am FB Informatik, Fachhochschule Darmstadt, 1997.

Part II

From CAD to VR

4 Data Preparation on Polygonal Basis

Christian Knöpfle
Fraunhofer-Institute for Computer Graphics (IGD)

Jens Schiefele
Technical University Darmstadt, Flight Mechanics & Control (FMRT)

Converting NURBS based CAD Data into polygonal data often leads to unconnected faces at patch borders, wrong sided face normals and holes in the calculated mesh. Several methods are described in this paper to overcome these problems. The tesselation also results in a huge amount of polygons. To conquer such large data sets, different decimation algorithms are described and compared. All these tools are embedded in the Delphi viewer. Delphi provides an interactive environment including scientific visualization for decimation previewing, parameter distribution, and vertex classification, all superimposed on the data sets.

4.1 Introduction

In industry, computer models of design objects are created in order to decrease the time and resources needed to develop a complete product. After an engineer has outlined his ideas about a new design it is constructed in a 2- or 3-D CAD System.

After a design object is constructed in a CAD system a prototype has to be manufactured to verify that all construction constraints were obeyed. In most cases this process has to be repeated several times because of design errors. Depending on the design object this can be an easy and cheap process (e.g. development of new screws in automotive industry) or extremely time consuming and expensive (complete new engine, car, etc.).

An alternative is Virtual Reality and Augmented Virtual Reality. The complete design object can be seen in 3D and explored with a data glove. A human user becomes part of a virtual scene. He can investigate ergonomics, verify construction constraints, rearrange virtual pieces, and test functionality while interacting with objects. These changes can be used to adapt the CAD model itself.

Augmented Reality offers the possibility to mix physical objects and virtual objects. This is helpful when only some objects have to be reconstructed while others are already existing.

Virtual Reality reduces cost and conserves resources by detecting design errors early and a fast evaluation/verfication of product characteristics. Therefore, it shortens design cycles and helps to create a better product.

CAD data sets are typically modelled with mathematical descriptions. These mathematical descriptions are in most cases given as Non-Uniform Rational B-Splines (NURBS). The advantage of a mathematical representation is that surfaces can be easily manipulated and changed. NURBS surfaces have usually a so-called continuity. For a human observer this is a smooth surface. Polygon meshes are only continously forming sharp edges. It is much easier to store NURBSs in smaller sections of memory and smaller data files. Usually a complete model is not made from one NURBS but an assembly of different NURBSs being attached to each other. Each such NURBS description is called a patch.

Todays computer graphics hardware is based on polygons as rendering primitive, not on NURBS. Therefore, it is necessary to convert CAD data from free form surfaces and solid models into a polygonal representation.
In this paper the process of converting free form based data models to polygon based data sets, which are suitable for Virtual Reality applications is described. All techniques explained in this paper are implemented in the Delphi-Viewer, which will be discussed too.

4.2 Converting CAD-data

The conversion of NURBS-based CAD Data to polygonal Data, the tesselation process, usually creates hundreds of thousands of polygons in order to provide a faithful representation of the model. With current computer hardware it is necessary to decimate these models down to about 10.000-80.000 polygons. This decimation is needed, because a scene has to be rendered at about 10-25 frames per sec for fully immersive model viewing, which is one of the major constraints in VR Environments.

Beside the huge amount of generated polygons, other problems are introduced caused by all available converters:

- NURBS-Patches doesn't form a closed suface
- small holes in the model
- wrong sided face normals

- coplanar faces
- non-simple geometry

For optimal decimation results and high quality rendering these problems have to be fixed.

4.2.1 Polygonal decimation versus controlled tesselation

For CAD data there are two different approaches to limit the number of polygons. Mathematical free form surfaces do not have a unique polygonal representation. Therefore, it is possible to assure during tessellation that the number of polygons does not exceed a certain number by selecting an appropriate tessellation. This method can only be used to limit the number of created polygons for free form surfaces. Unfortunately the problems like unstitched patches and wrong sided face normals still exist in todays converters.

Most CAD system also allow to define polygon based data sets. This data cannot be decimated without a polygonal decimation algorithm.

In addition, in most situations it is necessary to decimate polygonal data sets taken from other domains. Therefore, a more general polygonal based decimation is needed for most cases.

In this paper the second approach is discussed. A high quality tesselation, where a accurate polygonal model is generated, preprocessed and afterwards decimated to get a model with a handable complexity.

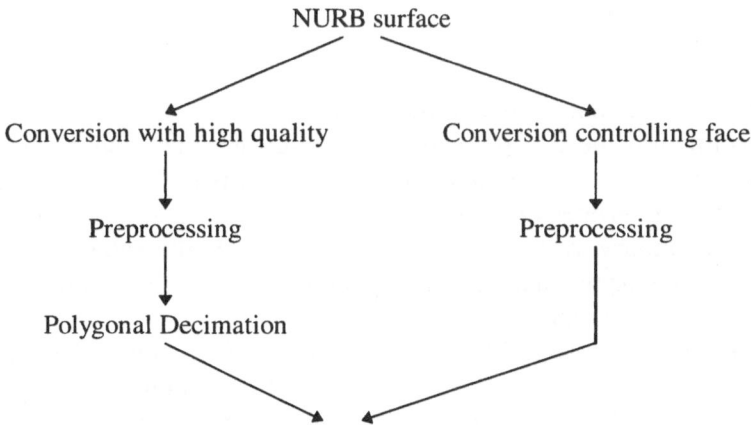

Figure 4.1. Alternatives for limiting polygon number of NURBS data.

4.3 Optimizing data for decimation and rendering

The problems introduced by converting NURBS-surfaces into polygonal models are described in this chapter.

4.3.1 Stitching

In most converted data sets boundary vertices of different patches do not coincide. Either the frequency of vertices is different on each side of the boundary, or they are simply on different positions. Therefore, boundary edges are also not parallel. They usually overlap or form small gaps along borders of adjacent patches. These gaps are immediately recognized by an

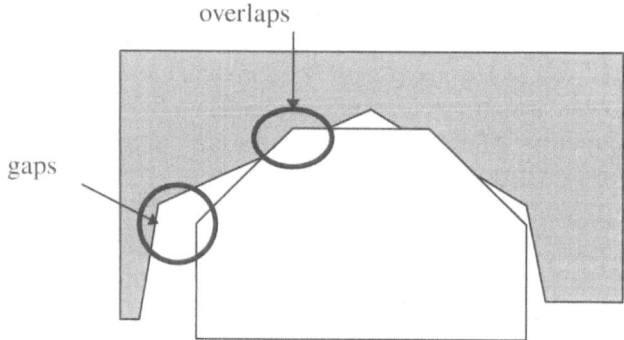

Figure 4.2. Unstitched NURBS patches

observer especially on distinct background colors.

During decimation these gaps tend to increase their size and become unacceptably large. In some decimation algorithms it is possible to maintain these boundary vertices of a patch, but decimation will lose effectiveness, because boundary vertices can't be decimated. Some models have overlapping patches or patches that are not connected at their boundaries but somewhere within a patch.

In [Fry95] a stitching algorithm was developed. It assumes that one patch after another is stitched to an existing mesh. First, boundary vertices of a patch closer to a boundary vertex of the mesh than a given limit are moved to the mesh point. This process is called „snapping". Second, for all unchanged boundary vertices the closest boundary edge of the mesh is calculated. While the distance to this edge is smaller than a given limit it „jumps" to the closest vertex forming this edge.

The concept of this algorithm was extended for more general use in [Schief96]. Three different algorithms can be chosen and executed in any combination:

- *Direct Snapping* merges all boundary vertices with their closest neighbour independent of surrounding geometry. Only a given distance must not be exceeded.
- *Vertex-Snapping* merges all boundary vertices with their next neighbor while checking that this neighbor is not connected with an edge and the distance is smaller than the closest connected neighbor.
- *Edge-Snapping* calculates the closest edge for all boundary vertices. If the vertices forming this edge are not edge connected with the vertex itself and the distance is smaller than a given limit, then the vertices are merged.

4.3.2 Filling

During a conversion process from NURBS surfaces to polygons small regular areas are left out. These areas are holes in an otherwise smooth surface.

Some CAD systems display objects only in a wireframe representation. A hole that is surrounded by faces can not be determined in wireframe. Therefore, a CAD designer cannot detect this design error [Luc95].

Converters do not check whether or not input polygons are planar. Computer systems can only render planar polygons because areas have to be uniquely defined. In order to solve this problem these non-planar polygons are removed and created holes are refilled.

Sometimes holes are created intentionally. In most cases these holes are formed by a long chain of vertices. Filling is usually executed only on a few vertices.

4.3.3 Face normals

With free form surfaces it is possible to calculate very precise vertex normals. Vertex Normals are used for Gouraud shading. In most cases they are lost during conversion because most converters do not support vertex normal calculation. Therefore, it is necessary to recalculate them based on existing face normals.

Face Normals are not used for rendering in CAD systems (flat shading). Hence, they are not taken into consideration during construction. However, in polygonal-rendered systems they are used for illumination and visibility calculation and recalculation of vertex normals. Their consistency is essential for good quality rendering.

There are several approaches one can generate consistent face normals. For smooth surfaces without manifold structures the problem of finding correct face normals is reduced to determining inside and outside of a given object. If one correct face normal is found, all others can be adapted accordingly.

Not all objects are physically correctly modeled and have a visible and invisible side. For example, a one walled half sphere does not have an inside or outside notion. Another method is by specifying a face and it's correct normal and then recursively search the attached faces. They are adapted by toggling faces with wrong sided face normals according to the specified face. This methods fails if the mesh has manifold or physically impossible topologies.

4.3.4 Coplanar faces

Identical-Coplanar faces are created during CAD construction process [Luc95]. In wireframe represenations it is not possible to detect double-defined surfaces. Sometimes, converters introduce double defined surfaces.

To elliminate coplanar faces polygons are checked for common vertices. Polygons with same vertices are removed and it is tried to preserve correct face normals by checking neighboring faces.

4.4 Polygonal decimation

Decimation reduces the number of polygons to a desired quality and quantity level. The main areas of application are:

- creating Level-of-Detail (LOD)
- using low detail approximations for illumination algorithms (radiosity) or intersection calculation (collision detection)
- simplifying overdefined models

A Level-of-Detail is only rendered when a certain criterion is fullfilled: distance from viewer, movement of object, size, or position in viewing field. The transission between two levels can be accomplished by switching, blending or morphing [Knö96]. Usually different LODs are generated to assure that there is in all situations an appropiate representation for rendering available. It is accepted that a LOD defined by low object complexity also has a low visual quality. However, even low level LODs should be the best possible representation for a given object quality.

Low level approximations used for illumination algorithms or intersection calculation have no need for high visual accuracy. They should approximate an object to the best possible quality with a very low object complexity.

CAD derived models are usually overdefined and need to be decimated without loss of quality but signifcant object complexity decimation.

Having this in mind, decimation can be defined as:

The capability of reducing the number of polygons and vertices to a minumum quantity while preserving as much visual quality as possible.

These goals can not be satisfied together. They are are almost in contradiction to each other. It is necessary to find an acceptable tradeoff depending on application and available hardware.

In general, there are four different schemes for decimation: Clustering, Data Removal, Surface Imitation, and Texture Mapping ([Schief96]). These schemes will be described and compared in the next sections.

4.4.1 Clustering

Every object is surrounded by a bounding volume. For clustering, this bounding volume of an object is divided into smaller separate sub-volumes. Each sub-volume is called a cluster. These clusters can be any geometrical object (e.g. cube, sphere, cylinder). It is only necessary to

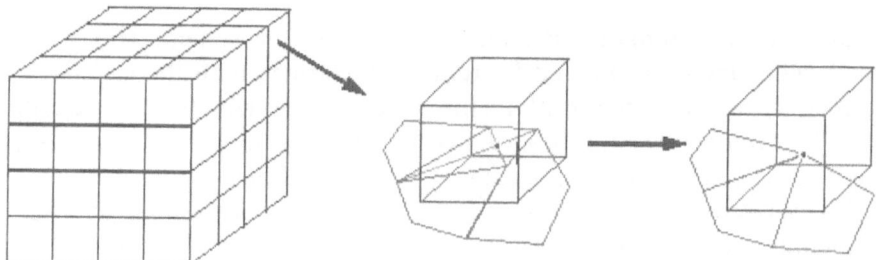

Figure 4.3. Principle of clustering algorithm

guarantee that all vertices can be clustered.

In most algorithms a bounding box with sub-cubes is used. In this case, the bounding box can be defined as the smallest box parallel to principal axes that includes all vertices of an object. The size of each cluster is determined by a user-specified parameter. Within a cluster vertices are merged together and a representative vertex is picked. In a second step all degenerated polygons are removed. A polygon is degenerated if there are less than two distinct vertices.

Depending on whether or not all vertices are chosen, how the clusters are calculated, and on pre-post processing steps clusterers can be divided into data sensitive and direct clustering. Clustering algorithms are described in [Rossi94], [Scha94] and [Schief96].

4.4.2 Data removal

From a given data set vertices, edges or polygons can be removed. Removal is based on algorithm dependent criteria: face angle, face size, face to average plane distance, edge length, edge angle, or vertex distances.

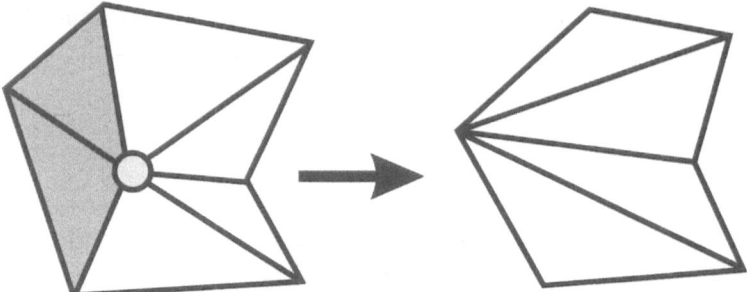

Figure 4.4. Principle of data removal algorithm

In most algorithms a triangulated geometry is assumed. This triangulated geometry is decimated by removing vertices with the criterion described above. If a criterion fails the vertex is kept without change, otherwise it is removed. After a vertex is removed the local area is retriangulated. Usually all vertices in a data set are checked in several runs.

Many decimation algorithms implemented in commercial as well as academic environments are based on [Schr92] average plane distance criterion. A vertex is removed when its distance to an average plane, calculated from all its neighbor vertices, is greater than an user-defined threshold.

Data Removal algorithms are described in [Schr92] and [Schr94].

4.4.3 Topological imitation

In contrast to the two schemes explained above, Topological Imitation does not necessarily use original geometry and topology to represent a data set ([Turk92], DeHa91]). Instead, a simpler geometry is calculated trying to imitate a given data as well as possible. In addition, this geometry has to consist of fewer polygons/vertices than the original.

4.4.4 Texture mapping

Polygonal decimation was defined as the removal of vertices/polygons while trying to avoid visual quality loss. Texture maps are an ideal decimation scheme when used from a distant viewpoint [Beig90]. A large number of polygons can be represented only by a few polygons to which the texture map is attached.

This method gives a very natural impression of a scene and is very useful. However, if a texture is very close to a viewer or observed from certain angles then it is possible to recognize flatness and distortion. It can only be used in some situations and is not a general decimation scheme.

4.4.5 Comparison of decimation schemes

Clustering is the most general approach for decimating all kind of polygonal data. It is independent of mesh connectivity. Even for decimation with high decimation rates the overall shape is preserved. In addition the error of decimation is bound by cluster size. Except for [Scha94], all algorithms are high performance implementations. They are the fastest algorithms available because almost no criterion evaluation calculation overhead is introduced.

The drawback of clustering is that it changes connectivity and destroys topology. It also tends to shrink objects. Unconnected faces can become neighbors after decimation. Smooth topologies are destroyed. Therefore, clustering is not very suitable for objects with tiny parts that have to be preserved. In Figure 4.6 two examples of a decimation using clustering algoritms are given. The original object is shown in Figure 4.5.

Figure 4.5. Arm with 7643 polygons

All data removal algorithms are excellent for preserving certain features. The kind of preserved feature depends on the used removal criterion. In data retrieved from CAD, areas of high curvature within a small area are preserved, while large areas are already flattened out. These small areas usually contain more than 50% of all polygons.

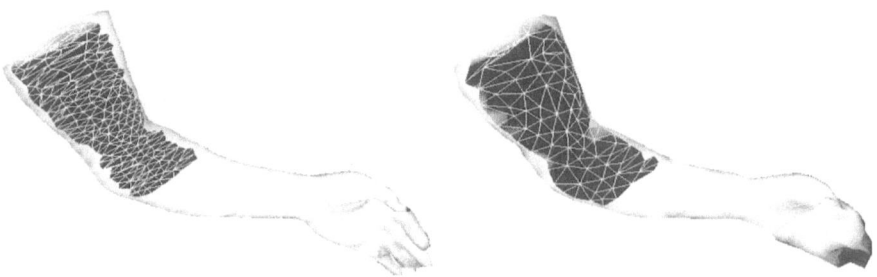

Figure 4.6. Clustering: Arm with 2750 and 828 polygons

For moderate decimation of equal sized polygons these methods are superior. They preserve desired regions and remove vertices/faces/polygons in flat regions.

An additional problem is „retriangulation dependence". Quality of decimation is essentially influenced on the used retriangulation algorithm. Most algorithms use very simple retriangulation schemes.

Most data removal algorithms are based on face angle [Schr94], or distance to face [Schr92] criteria. All these criteria fail for cylindrical equally sized objects. Therefore for CAD models of pipes, screws, and all kind of cylindrical objects, data removal algorithms are not usable. The arm in Figure 4.5 was also decimated using data removal algorithms (Figure 4.7).

Most topological imitation algorithms need smooth surfaces without sharp edges or corners. Eventually, existing corners or sharp edges will be immediately removed. However, they usually carry much visual information to users. Therefore, this decimation scheme is not very useful for most applications found in connection with CAD data.

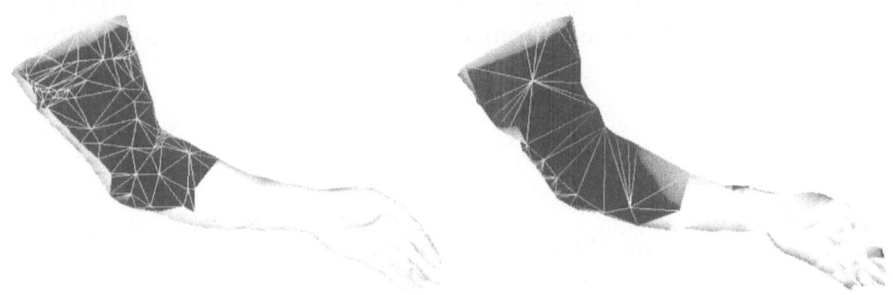

Figure 4.7. Data Removal: Arm with 2640 and 797 polygons

	Clustering	**Data Removal**	**Topological Imitation**
Timing	very fast	fast - slow	slow
Decimation	overdefined areas strict decimation	moderate decimation	all levels
Data Source	all kind of data	all kind of data	data without edges and corners
Advantages	bounded error preserve overall shape	preserve parts maximal user control	work for all levels good mathematical representation
Disadvantages	shrink model unsensitive removal	overpreserve parts destroy overall shape abstract parameters	not suitable for all data sources

Table 4.1. Comparison matrix of decimation schemes

For smooth surfaces decimation results are very good. The overall shape is preserved and data is usually removed evenly depending on removal criterion.

Usually best results were accomplished by combining a clustering algorithm with a Data Removal algorithm. For most applications [Rossi92] was alternated with Schroeders [Schr94] approach. First, all small overdefined surfaces were removed with a small parameter using [Rossi92]. In anintermediate step, data was flattened out with [Schr94]. Finally, [Rossi92] was run to calculate a final representation. This experience was used and included into a Automatic Decimation algorithm, described in [Schief96].

4.5 Delphi

Tools and decimation algorithms were embedded in a viewer to support the whole pipeline for decimation. This newly developed viewer is based on Y [Rein95], a high performance renderer, and XForms a general user interface builder. Convenience functions for displaying, viewing, and navigating were included to create a comfortable work environment. Work sets can be arbitrary selected. Data classification and inconsistencies can be visualized. For all algorithms a visualized distribution of decimation parameters and values is available. In addition a preview for decimation is included. All tools have a preview and statistical information. „Undo" can be executed on different levels for optimal decimation results.

4.5.1 Tools

As allready described in chapter 4.3, several problems are introduced during the tesselation process. Delphi provides several tools to conquer these problems:

- stitching (merges close boundary vertices together)
- filling (fills holes in a smooth surface)
- removal of dangling faces (removes faces that prevent a surface from being simple)
- removal of coplanar faces
- adjusting face normals

In addition, there are other tools to create a comfortable work environment.

4.5.2 Decimation

In Delphi several decimation algorithms are implemented. They can be directly started from the viewer. For all algorithms certain information and visualizations are available:

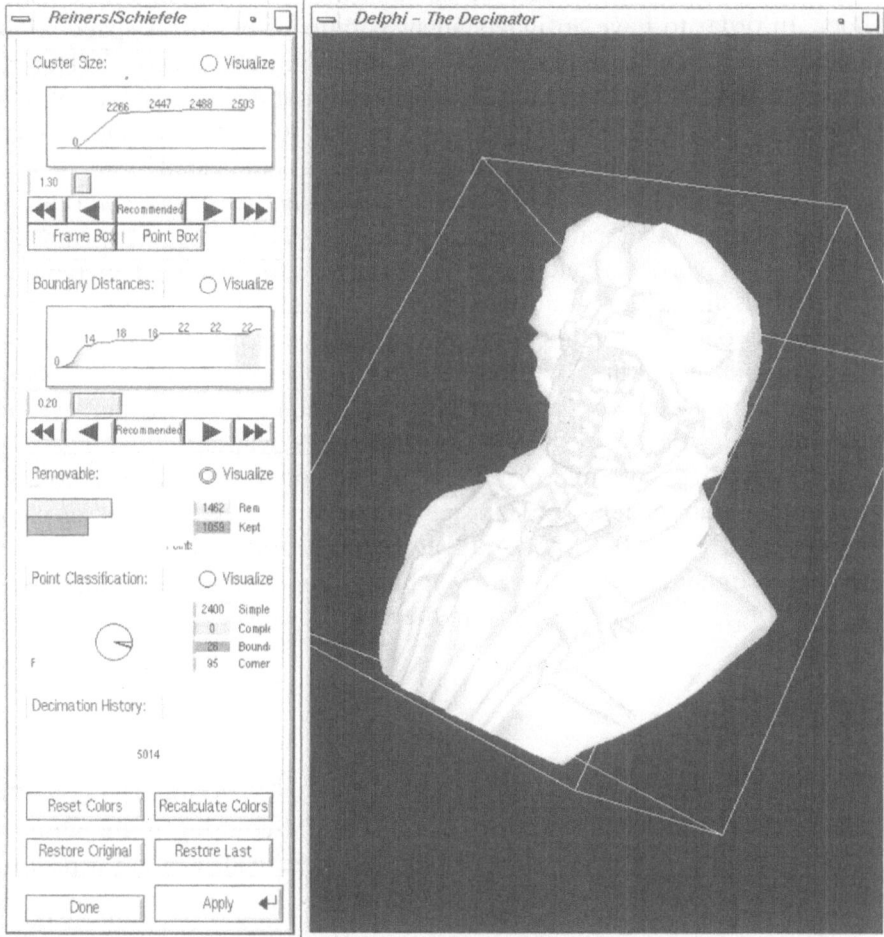

Figure 4.8. Different visualization methods in Delphi (removable vertices are displayed in red, see section Color Plates)

- data distribution concerning a selected parameter
- visualization methods superimposed on the geometry to show these parameters

- number of vertices to be removed
- visualization of removable vertices / polygons superimposed on the geometry
- visualization of vertex classification, according to the scheme described in [Schr92].
- number of polygons during decimation history

Each decimation parameter of an algorithm can be modified with a slider. In order to have an independent parameter of object space all parameters are scaled to a percent of the bounding box diagonal. This has proven to be an adequate value. In most cases parameters can be set independent of data type only based on visual quality to be expected. Above every slider, a parameter distribution of vertices is indicated.

In most applications it is of interest to see whether or not a certain parameter setting will create desired decimation. Therefore, a preview was included. All vertices candidates for removal are displayed in red. If a parameter is interactively changed it instantly adapts the data set. On the decimation panel the number of removable vertices is displayed. This visualization technique is usually used for decimation. Most people working with Delphi do not look for parameters anymore but use this visualization as the only criterion for parameter selection.

There are different „undo" notions implemented. It is possible to undo the last decimation step. This is usually done if a decimation did not have the desired effect. It helps to establish a certain 'trust' into Delphi. Due to the fast decimation, a trial and error approach for decimation is often used. Furthermore, it is possible to restore the original object.

4.6 Results

A car engine was taken directly from a Catia CAD system. The complete process of decimation except for conversion and coloring is supported in Delphi. Delphis interactive environment reduces needed time for decimation significantly.

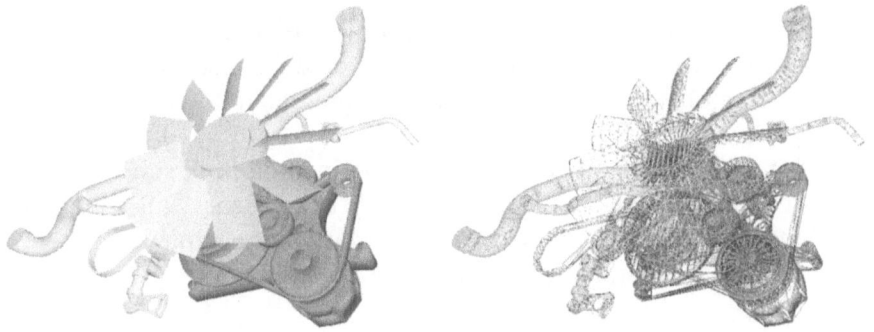

Figure 4.9. Front part of an engine with 71.345 polygons

The engine was converted to a polygonal model and preprocessed using Delphi (stitching, filling). The inital number of faces was 71.345. One single level of decimation was created. All objects were sequentely picked and decimation executed. In most cases only one run was necessary. For this particular data set clustering algorithms are superior because most objects have cylindrical parts or shapes. All other algorithms tend to fail. Face angles about 37 degress for [Schr94] and cluster sizes between 0.6 and 2.8 Percent for Data Sensitive Clustering ([Schief96]) were used. 2 hours of work time were needed and the result are 24.542 faces. These are 34% of the original faces or an overall decimation of 66%. The work time includes loading, storing, restoring, decimating, changing of parameters, taking notes etc. and was measured with a stop watch.

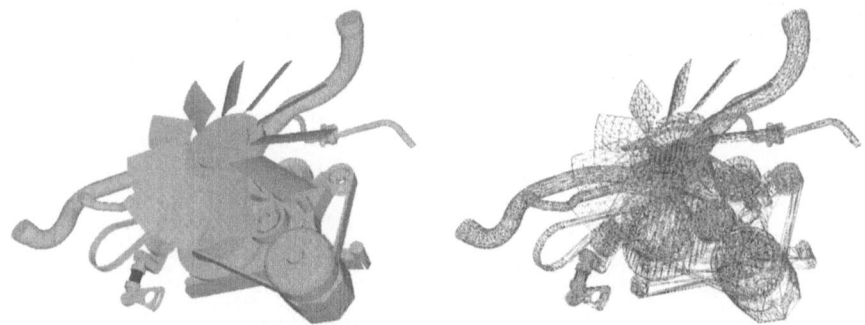

Figure 4.10. Front part of an engine with 24.542 polygons

A belt driven pulley taken from the engine is shown in Figure 4.11. It is decimated by 82 percent. In this case, decimation removes all small parts but preserves exactly the overall shape.

As a second example a pipe is displayed in Figure 4.12. It was decimated by 75 percent without loss of quality. In this case the result is not a generalization of an object but the same object representation with fewer polygons.

The fan and the belt of the engine are defined on both sides. The distance between front and rear part are very small. Hence, all decimation algorithms cause sides to intersect and overlap after decimation. There is no algorithm that consideres the complete topology.

Figure 4.11. Belt driven pulley: decimated from 2956 faces down to 524 faces

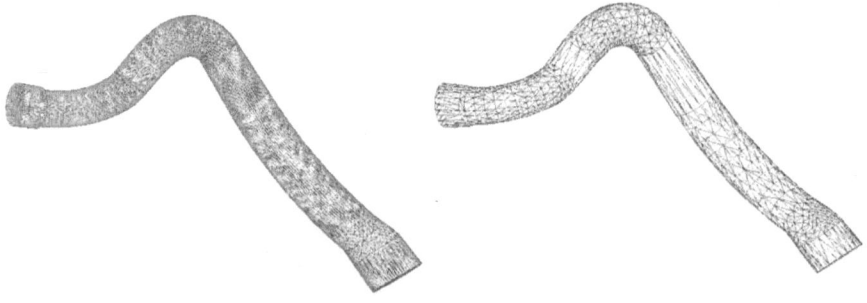

Figure 4.12. Pipe: decimated from 6168 faces down to 1588 faces

4.7 Conclusion

In this paper we presented a way to convert CAD-data into a polygon-based model, which satisfy the needs for a VR application. This approach includes a high quality tesselation, preprocessing and polygonal decimation. We also outlined why this approach is superior to a 'controlled tesselation' method.

The tesselation step introduces many problems to the database, like unconnected vertices at patch borders and inconsistend face normals. To overcome these problems and to optimize the data for the decimation step, several tools were implemented in an interactive viewing environment, called the Delphi viewer.

It is necessary to mention that the complete preprocessing-step is only necessary because today's converters cannot create a high quality tessellation of NURBS patches. All problems mentioned above can be solved with a conversion algorithm that uses the available information on the level of the mathematical description.

We also presented a classification of polygon-based decimation algorithms. Their main principles are outlined and advantages and disadvantages are described. Finally they are compared to each other.

Algorithms of each class are included in the Delphi viewer. All decimation algorithms have a preview and visualize valuable decimation information. It is possible to examine topologies and find design and conversion errors simply by their color-coding. Finally, the capabilities of Delphi to support the process of virtual prototyping were demonstrated on an engine.

Reference

[Beig90] Beigbeder J., Jahami G., Managing Levels of Detail with Textured Polygons, Ecole des Mines de Saint-Etienne, Departement Informatique, France

[DeHa91] DeHaemer M., Zyda M., Simplification of Objects Rendered by Polygonal Approximations, Computer&Graphics, Vol.15, No.2, pp.174-185

[Eck95] Eck M., Tony DeRose et al. Multiresolution of Arbitrary Meshes, SIGGRAPH95, pp.173-182

[Fol90] Foley J., vanDam A., Computer Graphics and Principles, Addision Wesley, 1990

[Knö96] Knöpfle C., 3D-Morphing and it's Application to Virtual Reality in Göbel M., Virtual Environments and Scientific Visualization '96, Springer, 1996

[Luc95] Luckas V.,Topologie Analysen von Geometriedaten und deren adaptive Aufbereitung für die Visualisierung, Diploma Thesis, Technical University Darmstadt, IGD, 1995

[Fry95] Fry K., Decimation of Triangle Meshes with Feature Retention, Technical Report, UofM

[Rein95] Reiners D., High Quality Rendering for Virtual Environments, Diploma Thesis, Technical University Darmstadt, IGD, 1995

[Rossi92] Rossignac J., Borrel P., Multi-Resolution 3D Approximations for Rendering Complex Scenes, IBM-Research Division, Research Report RC17697(#77951)

[Scha94] Schaufler G., Stürzlinger W., Generating Multiple Levels of Detail from Polygonal Geometry Models, Proceedings of 2nd Eurographics Workshop on Virtual Environments, Monte Carlo, pp. 53-62

[Schief96] Schiefele J., Methoden der automatischen Komplexitätsreduktion zur effizienten Darstellung von CAD-Modellen, Diploma Thesis, Technical University Darmstadt, IGD, UofM, 1996

[Schr92] Schroeder W. et al., Decimation of Triangle Meshes, SIGGRAPH92, pp.65-69

[Schr94] Schröder F., Roßbach P., Managing the Complexity of Digital Terrain Models, Computer&Graphics, Vol.18, No.6, pp.775-783

[Turk92] Turk G., Re-Tiling Polygonal Surfaces, SIGGRAPH92, pp.55-64

5 SIMPLIFY - Application Specific Data Preparation

Franz-Michael Hagemann
Audi AG, Ingolstadt, Germany

A long term goal of product design is to directly go from a virtual model to the production. Models created with today's CAD tools, usually are surface/face models. Additionally, the models become more and more complete and detailed, including all rounding and facets. Due to this, huge amounts of data are created along the planning and the design phase. The more complete the description of the CAD parts is, the more data must by handled. What on one side must be rated as a benefit (a complete and with that a clear description of the part), becomes on the other side a disadvantage, if one likes to do calculations or simulations with these data. To reduce the calculation time to a justifiable extent, simplifying assumptions about the geometric shape of the parts are necessary and permissible. Due to the fact of large amounts of data created for each car and its variants in the automotive industry, a manual simplification of the data is not sensible. An approach for automatic data preparation is introduced in this article, a prototype will be described.

5.1 Introduction

Description of the CATIA Module SIMPLIFY

The size of CAD models increases significantly. Product components, fixtures and tools become more and more detailed to fulfill the requirements of the process chains. This leads to problems when CAD data are to be reused in special applications. For example the visualization speed of simulation systems slows down depending on the amount of data to be displayed. In digital mock-up tasks an easier handling is possible with less data.

This makes a processing of the CAD models necessary according to the requirements of the specific applications. In principle three different procedures can be carried out with SIMPLIFY:

1. A *reduction* of the size of the CAD models to the required level for the specific application. This means to remove all data in the model which do not belong to the part description (application specific data processing).
2. A *simplification* of the CAD model by using techniques to replace complex entities with more simple entities.
3. An increase of the *data quality* by repairing inaccurate elements in a CAD model. This means an improvement of the data quality (mainly with respect to its mathematical description, e.g. to replace multiple patches by one surface)

The CATIA application SIMPLIFY is designed to carry out these three procedures.

The benefits that can be reached are a significant reduction of the model size (50% to 80% possible) and with this a better handling in the down stream process chain.

5.2 Description of the problem

Most of the product parts and plant facilities in the automotive industries are planned and realized with assistance of CAD tools today. These CAD models become more and more detailed, as it is necessary to manufacture the corresponding parts. A lot of information is entered into the CAD models to be used in later steps in the process chain. But some specific applications do not need all these detailed data. On the contrary the high level of detailisation and additional information affects the visualization speed in robot simulation tasks for example.

Due to the fact that the computers now being used in conjunction with CAD are becoming increasingly powerful, the users are also coming to expect faster processing and more convenient operation. In particular, the problem of speed can only partly be solved by the introduction of workstations optimized for graphic data processing.

Whereas a few years ago it was normal practice to use only wireframe models for displaying objects and design work, now surfaces and solids are being increasingly used for depicting and working on objects.

The reasons for the use of enhanced entities are:

- display with hidden lines and surface shading
- Boolean operations
- Physical properties
- Simulation for collision investigations

- Exact mathematical description of geometry
- Avoidance of design errors.

As a consequence of the increased use of surfaces and solids the requirement for memory capacity will expand further. At present a model comprises about 4 - 8 megabytes, and larger models are already anticipated.

The reason for the large memory capacity requirement is the degree of accuracy of the depiction. The more facets required for depiction, the greater the degree of accuracy and the greater the memory capacity required for the entity.

The purpose of SIMPLIFY is derived from this problem: to simplify surfaces and solids in 3D models with the CATIA CAD system.

5.3 Solution to the problem

5.3.1 Reduction of the size of CAD models

An internal investigation has demonstrated that a large part of the planning work is used to clean the CAD models according to the specific requirements of the different applications.

Reasons for this partly considerable post editing are:

- different ways of creating data in various CAD systems
- different internal data representations of the CAD systems and
- miscellaneous tasks of the data producer and the data receiver

To solve this problem the CATIA module SIMPLIFY was enhanced with a functionality to process and clean a CATIA model. This module deletes all unnecessary geometric elements (like boundaries of faces and surfaces, auxviews, layer filters, 3D text, unused details and symbols). The CAD model is reduced to the "part describing" data (see Figure 5.1).

Additionally the user can move surfaces and wireframe data to NOSHOW. Furthermore an adjustment to company specific model standards can be carried out (see Figure 5.2, SIMPLIFY MODEL, cleaned model).

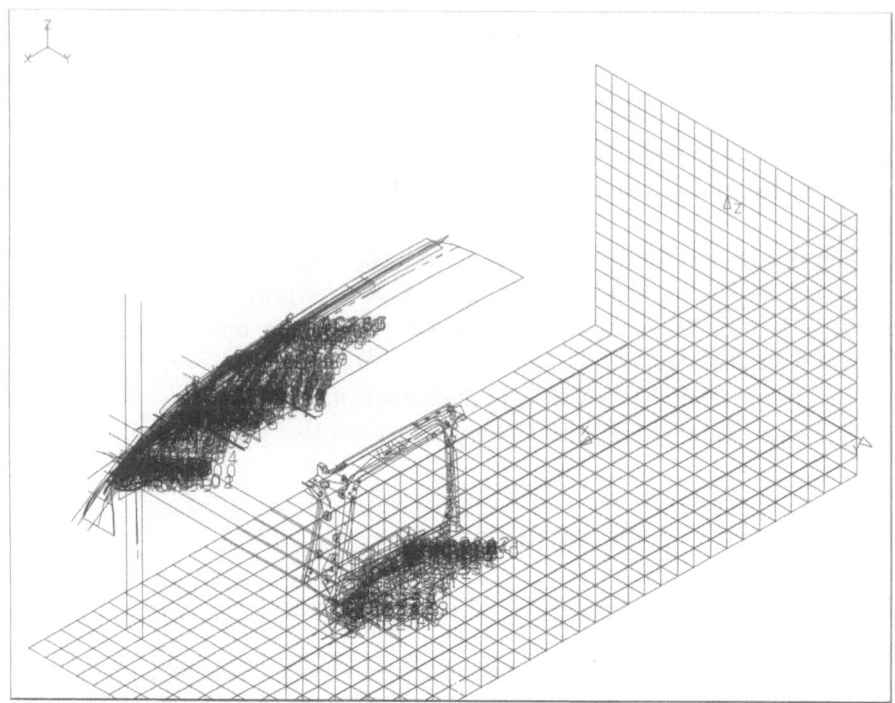

Figure 5.1. SIMPLIFY MODEL, original model

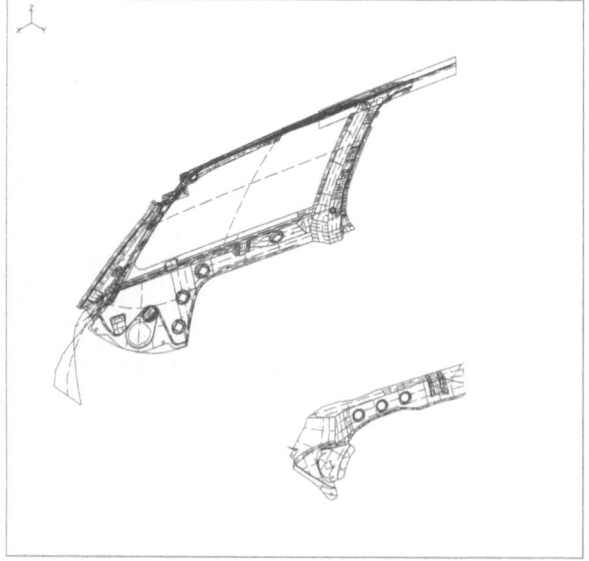

Figure 5.2. SIMPLIFY MODEL, cleaned model (Face geometry only)

5.3.2 Simplification of the CAD model

As a first approach the possibility of generally replacing entities with convex hulls was rejected, because the level of simplification was too high.

On the basis of these results it was decided to offer a selection of several simplification strategies in the application that was to be programmed. The common factor is that simplification is achieved by reducing the number of facets in the model.

The simplification strategies are:

- clean the model from unnecessary entities
- Reduction of discretisation for rotation bodies and curves
- Generation of enveloping bodies which wrap the solid in an optimum fashion
- Generation of convex hulls with minimum volume
- Elimination of redundancies in the model.
- reduce the mathematical complexity of a CAD model

While the first strategy does not alter the geometry of the model, the three other strategies involve manipulating the model: either solids are removed, or the enveloping bodies have to replace the removed solids.

Which of the strategies offered by the program is used is decided by the user on an interactive basis before simplifying the model. Here, among other things, the user can decide how many facets are to be used for curves and rotating bodies, or what type of enveloping body for certain types of solids. The user also decides at this point which solids can be simplified and which may not. Certain types of solids (cylinders, spheres, etc.) and certain specific solids can be excluded from the simplification process. In the application we have referred to this as INCLUDE or EXCLUDE.

Once the strategy has been decided, the model can be simplified. Here we have also provided several alternatives from which the user can choose:

- Simplification by batch program
- Simplification of a solid by selection with mouse
- Simplification of several solids of the model by multiselect

The selected solids are then simplified once the user has confirmed his selection. The illustration shows the simplification of a robot welding tool.

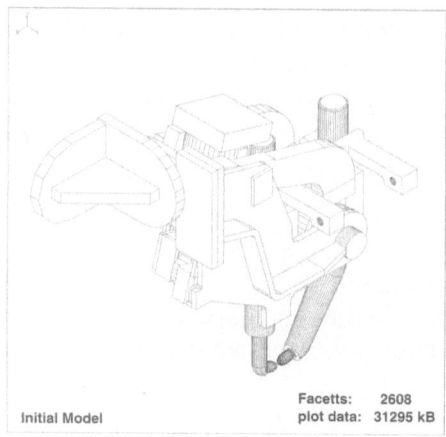

Initial Model   Facetts: 2608
                  plot data: 31295 kB

Figure 5.3. SIMPLIFY MODEL, original model

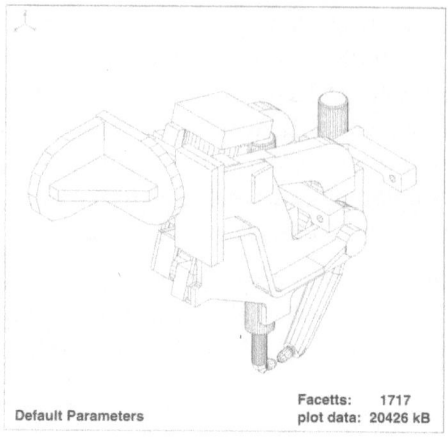

Default Parameters   Facetts: 1717
                       plot data: 20426 kB

Figure 5.4. SIMPLIFY MODEL, result with default parameters

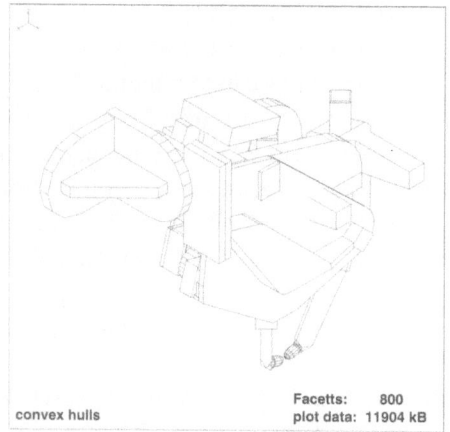

convex hulls   Facetts: 800
                 plot data: 11904 kB

Figure 5.5. SIMPLIFY MODEL, convex hulls

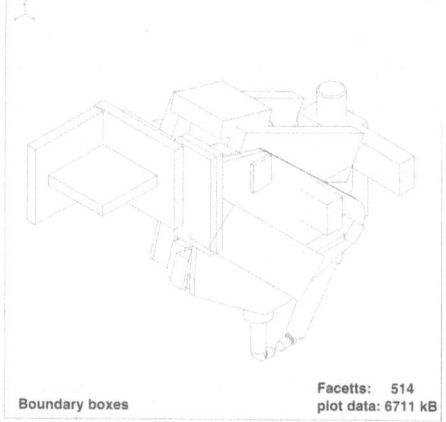

Boundary boxes   Facetts: 514
                   plot data: 6711 kB

Figure 5.6. SIMPLIFY MODEL, boundary boxes

The solids are replaced by their enveloping bodies as far as this was possible. In this case the original solids are removed and replaced by other solids.

5.3.3 Data quality of CAD models

In some cases surfaces consist of many small patches. With SIMPLIFY it is possible to approximate the multi-patch surface by one (see Figure 5.7 and Figure 5.8).

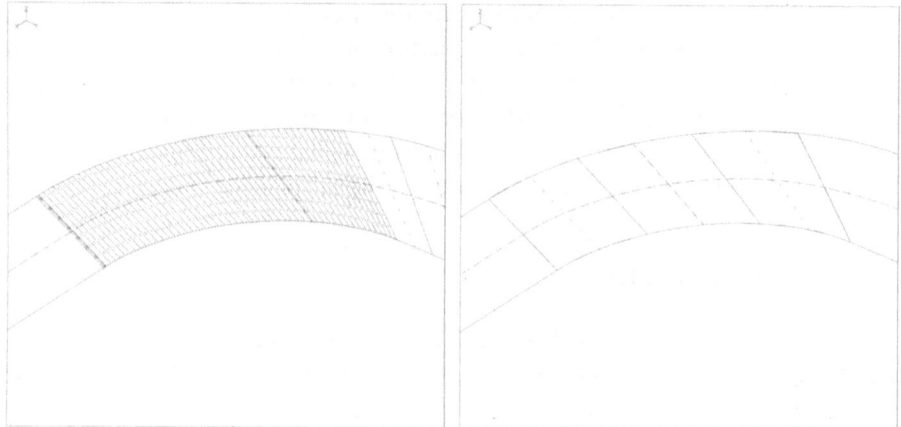

Figure 5.7. Surface with Multi patch   **Figure 5.8.** Approximated surface

5.4 Summary

5.4.1 Results

The use of the application on the depicted models shows that about 50% memory capacity can be saved and the simulation speed can be increased by a factor of 2. These figures depend a great deal on the type of elements used. The degree of simplification is also relatively important because the number of facets in the model ultimately determines what effects can be expected on the size of the model and the simulation speed.

5.4.2 Problems

No problems are normally encountered with the simplification of solids in models. However, if solids have been linked together in the design in such a way that they only have minimum intersection volumes, for instance, this can lead to problems when simplification is performed: after a change in the discretisation CATIA can no longer perform an UPDATE because the calculated intersect volume is too small.

5.4.3 Outlook

A consideration is that simplification is also useful for other applications. Since the same arguments regarding memory capacity apply here. It seems a good idea to modify the menu structure of SIMPLIFY to just offer the user a function like "SIMPLIFY for APPL_NAME", where APPL_NAME means a special application like robot simulation, NC programming or crash calculating. In this case all parameters necessary for the specific simplification process are implemented in the software.

Acknowledgement

The author wants to thank Mr. Schiffler, in those days working at his diploma thesis at AUDI AG, Prof. Dr. Pöschl freelance contributor at AUDI AG, and as well as Mr. Reinhard Kreitmayer and Mr. Manfred See, both colleagues and employees of AUDI AG, for their active cooperation to realize the SIMPLIFY program and their engagement, their belief in the feasibility of the program.

Part III

Applications

6 Virtual Reality (VR) - New Methods for Improving and Accelerating Vehicle Development

F. Purschke, R. Rabätje, M. Schulze, A. Starke, M. Symietz,
P. Zimmermann
Volkswagen AG; Corporate Research, Wolfsburg, Germany

Virtual Reality (VR), until recently the domain of the research labs, is starting to be accepted as a general business tool. VR allows users to see and explore new products, plans or concepts long before they exist in reality, in a continually more realistic manner, which at the moment no other technology can offer.

VR is a suite of 3D graphics, simulation tools and technologies which allows users to operate within a computer-generated environment, on an interactive basis and therefore in real time. VR in the automotive industry offers the possibility to enhance the quality of the product, as well as its time- and cost effectiveness.

In a research project which started in 1994 in cooperation with the Fraunhofer Institute IGD in Darmstadt, Germany, the possibilities of using VR in the area of vehicle development, manufacturing planning and marketing were investigated.

Today, after 2 years, the first applications are already being used successfully in different development areas. Further interesting applications will follow as soon as the technical prerequisites are available.

This paper provides an overview of the status of VR as well as the motivation which exists to incorporate this technology into the vehicle development process. Additionally, in taking some examples from the practical point of view, the successes and difficulties encountered are presented in more detail.

6.1 Introduction

What are the Motivations for an Automobile Manufacturer like VOLKSWAGEN to deal with Virtual Reality Technology and to Incorporate it into the Development Process ?
The development of a new vehicle costs not only a great deal of money, it also takes a long time. From the first drawing in styling, through through

the creation and approval of the product definitions until start of regular production, it lasts approximately 4 years.

Within this timeframe many tasks must be performed in parallel. For example, tooling must already be planned and manufactured. Engineering and development, design and test activities must take place. Simultaneously, numerous vendors must plan and develop their contributions. Before the first prototypes are manufactured, important product decisions have to be made.

In recent years, in order that individual areas of expert responsibility may support each other better, teamwork between different departments has been incorporated more and more. Simultaneous Engineering Teams (SET's) have therefore been formed. Among their responsibilities are - for example - investigations regarding matters with front compartment layouts, ergonomics, assembly-/ disassembly-, investigations for manufacturing and service as well as comparisons between different versions, functional investigations, manufacturing progressions and much more. A portion of these tasks can be accomplished in a more expeditious, cost effective manner and thus may be resolved more effectively through VR technology.

6.2 The VR project at Volkswagen research

The VR project at VOLKSWAGEN Research started in the beginning of 1994 with the goal of investigating the applicability and suitability of VR technology for the development process. At this time, only scientific institutes, universities and a few companies were involved in VR.

It was VOLKSWAGEN's desire to become involved in the application part of VR, therefore a partnership was established with the Fraunhofer Institute of Computer Graphics IGD, located in Darmstadt, Germany, which is a leading pioneer in the field of VR technology.

IGD supplied the basic software, which consists of three main packages:

- Rendering
- Interaction
- Collision Detection

In the course of the cooperative work effort, interactive techniques were developed, whose necessity became apparent during the targeted application framework at Volkswagen.

6.2.1 VR - the three-dimensional human-machine-interface

Virtual Reality is a three-dimensional Human-Machine-Interface (HMI) which allows the user to „dive" into and operate in artificial, virtual surroundings. The more perfectly his senses are appealed to, the better and more comfortably he will be able to situate himself in this „world".

The bandwidth of VR applications is enormous. It stretches from surfing the internet on a PC to three-dimensional „walks" through buildings and finally to complex and highly interactive applications from the world of technology and science as, for example, the „virtual windtunnel" or astronaut training at NASA.

Depending upon the nature of the situation, demands upon the hardware are very different. While PC performance is sufficient for one task, even the newest super graphics computer like Silicon Graphics' Onyx Infinite Reality is not even fast enough for others. Also, the interactive devices which are necessary to operate in cyberspace are in no way sufficient to meet the requirements for many applications in industry and science. There are for example severe deficits regarding haptic feedback.

6.2.2 What are the components of a professional VR environment ?

First of all, a high performance graphics computer equipped with a visual stereo output option is necessary. At the moment, the most commonly used hardware is comprised of models from SGI like the Indigo MAXIMUM IMPACT or the Onyx Infinite Reality with one or more graphic pipes.

Furthermore, stereoscopic displays are required. Here the degree of required „immersion" determines which stereo display is necessary. A „Head Mounted Display" (HMD) for example, secludes the observer completely from the outside world (higher/ increased degree of immersion, one's own presence shown only via graphical echo of, for example, the hand).

The stereo monitor, which together with shutter glasses creates a stereoscopic view for the user, has a low degree of immersion but a higher degree of presence, as the observer can see his own body in reality during the experiment.

There are also combinations of these extreme forms, for example the stereo wall, the workbench or the most complex device, the CAVE. The CAVE is a multiside stereo projection system for active glasses and has the advantage of high immersion and high presence. The disadvantages are an expensive price, the huge amount of space and the bad light intensity and contrast as compared to the stereo wall.

In order to interact with the artificial environment input devices are necessary. In the most simple case this can be the well known 2D-mouse, a 3D-mouse (with 6 DOF's) or even more sophisticated and complex, a data glove with 18 joints. Speech recognition is another favourable VR input device because it support a natural form of human interaction.

For positioning in space, trackers are required which determine the coordinates and angles of orientation. The most common today are work with magnetic fields. Unfortunately they are very sensitive to ferrons and not very accurate. In the future, optical systems will become more and more common as they work with a higher degree of accuracy (disadvantage : line of sight has to be preserved).

Speech recognition is very suitable for interaction in conjunction with acoustical feedback because in comparison to the normal computer user the cybernaut is unable to work with a keyboard or mouse.

Force- and tactile feedback are natural components of human interaction. These feedbacks are therefore very important for assembly simulations but also in evaluating shapes and surfaces in cyberspace. The development of such devices is however in its early stages. For this reason, attempts are being made today with other methods (acoustic feedback, autonomous virtual agents) in order to avoid these problems.

What kind of hardware configuration is Volkswagen using today and will use in the future for its Virtual Reality Lab (VRLab)?

For professional use in industry, the requirements for VR hardware are quite extreme. In spite of intensive and intelligent data reduction, including techniques like Level-of-Detail and Texturing, the number of polygons for the rendering of a scene e.g. the front compartment including engine and other units, can exceed 500 k.

With 10 frames/s stereo images, the required performance would exceed 10 million polygons/s. This number is approx. 5 times higher than the performance of the most powerful graphics computer today.

The following table shows the current VR equipment of the VRLab. shows the network of the VR hardware. The distribution between the various computers and the other equipment is highly flexible and cannot be shown clearly in the figure.

No.		Device	Used for
1.	1	Onyx Infinite Reality 8 x R10000 processors, 2 GB memory, 2 graphic pipes, 20 GB disc space	Rendering, collision detection for HMD and Stereo Projection
2.	1	Onyx Infinite Reality 8 x R10000 processors, 2 GB memory 3 graphic pipes, 20 GB disc space	Rendering, collision detection for HMD, Stereo Projection and for the near future: CAVE
3.	1	Maximum Impact 1 x R10000 processor, 256 MB memory, 1 graphic pipe	Testing & developing VR software Rendering, collision detection stereoscopic screen
4.	3	Solid IMPACT Workstations	Preparing data, software development
5.	3	Data Gloves	Interaction
6.	3	Tracker Polhemus Fastrak/ Flock of Birds	Tracking head & hands
7.	8	3D-mice	Navigation, interaction
8.	2	Speech recognition system	Speech input via microphone
9.	2	Audio system	Acoustic feedback
10.	1	Head Mounted Display n-Vision, High Resolution	Stereoscopic display with high resolution up to 1280x1024 interlaced
11.	2	Stereo Projection System TAN passive polarisation, 2 Beamer	Stereoscopic display with 2/ 3 m diagonal
12.	1	CAVE multi stereo projection (planned)	3 or more sided active stereo display system with shutter glasses

Table 6.1. VRLab/ VRStudio Hardware

What are the software components necessary to work within a virtual environment ?

Ideally, all the data needed to deal within a virtual environment are CAD data which already exist as a complete set and are already prepared by the CAD system for use in VR.

At the moment we are far away from this state. In reality we have the following problems :

- the product is not completely described by CAD data
- the CAD system normally does not prepare data in an ideal way, reduced and tesselated
- the interfaces like IGES only contain geometric information (unlike STEP)

Some CAD systems are now beginning to supply tesselated (which means conversion from the parametric description to a polygonal representation) parts in form of polygonal formats like INVENTOR or VRML. Nevertheless the quality achieved by an automatic tool does not reach the one supplied by a human. There are a lot of difficulties to overcome in a surface description, like for example the problem of t-boundaries, where so-called gaps are produced.

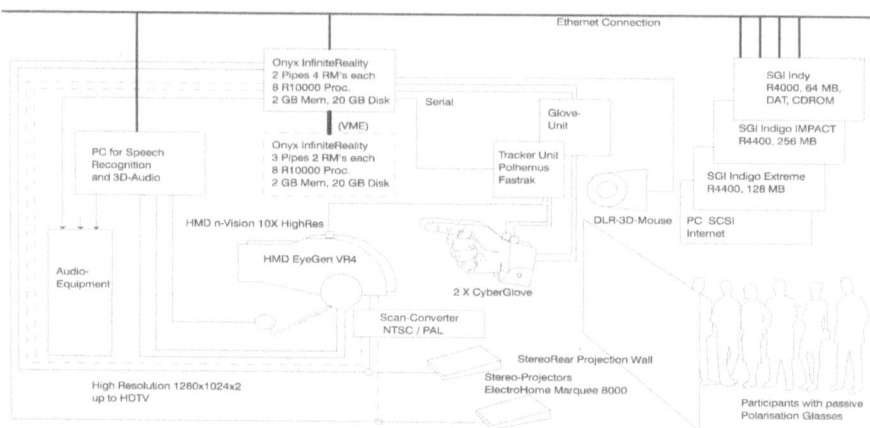

Figure 6.1. Hardware Configuration at VOLKSWAGEN VRLab

T-boundaries quite frequently occur from the surface design in CAD programs.

For high quality rendering in a VR environment they have to be closed or eliminated during the conversion- and reduction process.

At VOLKWAGEN VRLab there are software tools which are used to convert and reduce CAD data in different ways. Depending on the interface and the future quality requirements the data are prepared in different ways for use in the VR environment.

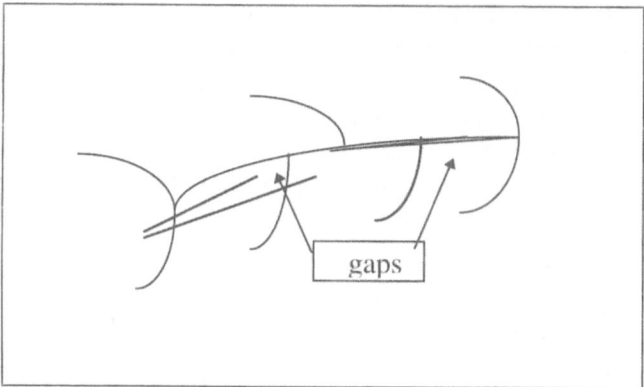

Figure 6.2. Gaps from t-boundaries

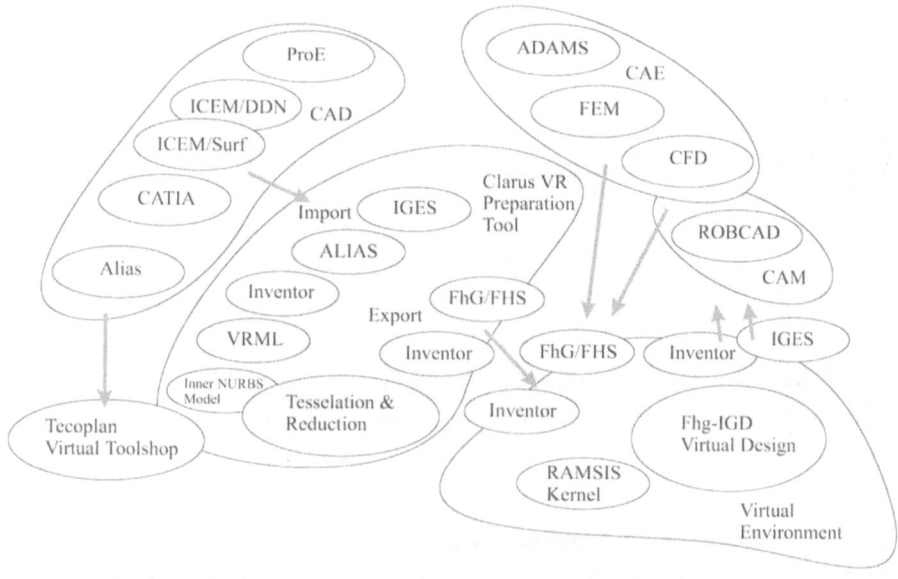

Figure 6.3. Data Conversion and Preparation for the VR Environment

Data reduction is necessary in order to achieve a sufficient frame rate during the real time rendering process. Real time is an essential requirement in VR and is itself dependent on the immersion and interaction requirements of the different „walks".

Figure 6.3 shows the process path from CAD to VR with the various interfaces, import and export.

The „walk" itself (this expression is derived from the applications where somebody is walking through a building in a 3D environment) is a highly sophisticated and complex real time process sometimes with a graet deal of interaction and collision detection as well as the rendering of many polygons. These tasks are preferably performed on multi processor computers with the different processes running in parallel (Figure 6.4).

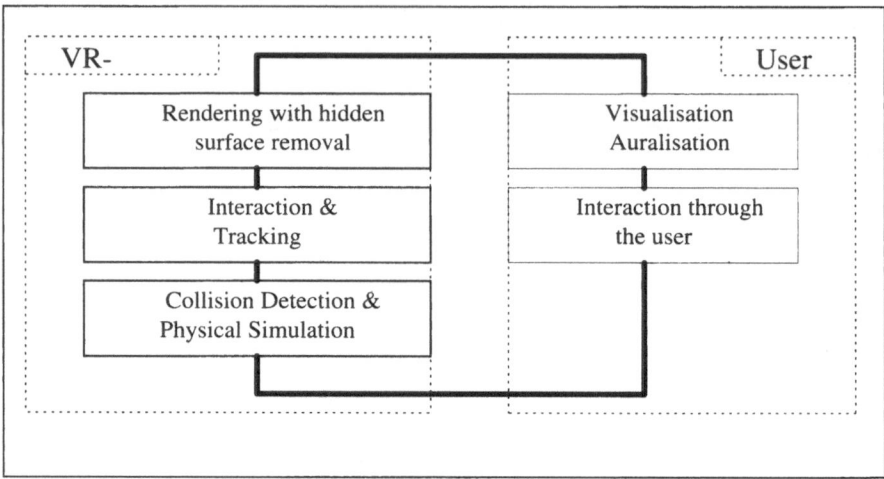

Figure 6.4. VR Environment Closed Loop

The schematic simulation loop in a virtual environment.
The real time requirements depend highly on the immersiveness of the application.

6.2.3 Examples of applications which have already been incorporated in the development process or which will be incorporated in the near future.

Coupling of the ergonomic dummy „RAMSIS" with VR

Ramsis is a product developed by the German vendor Tecmath by request of the VDA (Association of the German Automobile Industry). The ergonomic dummy Ramsis offers the possibility to obtain objective statements relating to a driver's behaviour (e.g. arm and leg positions, view, comfort).

By placing Ramsis into the VR environment it is possible to move the dummy around as well as to place oneself inside the dummy as an individual in any chosen visual and seating position.

Some examples will demonstrate this:

- The ergonomics team is investigating seat relationships in a new vehicle concept. In addition, various dummy sizes (5% female up to 95% male) are placed on the seat in order to cover the total spectrum of possible drivers. The dummy takes the most probable position by itself, which is controlled by the underlying software. The engineers can now observe for example on the stereo wall if the space remaining for the feet is sufficient, if the knee of a large man strikes against the steering wheel or if the side of the head is provided with enough room in the uppermost area of the roof frame.
- The observer becomes part of the dummy and can see whether the visual conditions are sufficient in front of a traffic light or if - for example - the sun-visor can block the view.
- The ergonomics team checks which parts of the instrument panel are covered by a cone representing the area of best vision. This vision cone defines the area in which a driver can most accurately see without having to move his head. This is important, since the most important in-

Figure 6.5. Using the ergonomic dummy RAMSIS in VR (see section Color Plates).

struments should be placed within the area defined by this cone.

• The dummy is given the task of sitting in a leaning position towards the rear of the vehicle in order to examine the posotion of the so-called C-pillar. In order to accomplish this, the dummy first of all has to be properly positioned. This task is of high kinematic complexity since many parts of the body must move relative to each other in a determined sequence (head, upper body, arms, even the legs). The quality of such software can by judged by the detail achieved in positioning the dummy: poor models with a smaller degree of freedom assume a robot-like position. After final positioning of the dummy the observer himself may again become a part of the dummy and see the dummy's view in order to evaluate with his own eyes if the area viewed may be deemed sufficient.

Virtual models as a replacement for physical models for surface inspection

The costs of building clay models and prototypes during the vehicle development process are enormous. Very often there are many iteration steps necessary before a whole model or parts of it are given the green light. If it were possible to deliver a certain amount of these models in VR in the high quality required, a great amount of both money and development time could be saved.

A milled data check model for a complete vehicle costs, for example, between 3 and 4 thousand work hours and thus approx. half a million DM. Even a smaller part model can only be delivered within 3 - 4 weeks and costs several ten thousand DM.

Virtual models not only have to show shape and surface in high quality derived from CAD data but should also offer some physical features such as shininess (highlights, colours, reflections), contour changes (flexible parts) and functional features such as the movement of a sun visor or opening and closing of a door. Interesting and important are also the possibilities to indicate the influence of tolerances on the image when being viewed.

Some examples may illustrate the advantages:

• The consequences of the tolerances in the upper hinge point of the driver's door are shown relative to the overall gap configuration in the vehicle. It can be seen that the 2 mm sinking in the rear area is remarkably noticeable from a certain viewing position.

• A designer investigates the progression of the highlights on top of the vehicle's roof with the help of virtual neon tubes which are mounted on the ceiling. In order to accomplish this, he stands in front of the stereo wall. The movements of his head are tracked by the computer. If he moves, his new viewing position is computed in real time and he can see the movement of the highlight, as if it were real.

- The gap between the instrument panel and the inner door is to be evaluated from the driver's position, since it is critical. In this case, in addition to normal rendering, a static radiosity calculation is performed in order to increase the realism by supplying the global illumination effects. This is necessary n addition to the normal shading because the influence of shadows are very important in this case.

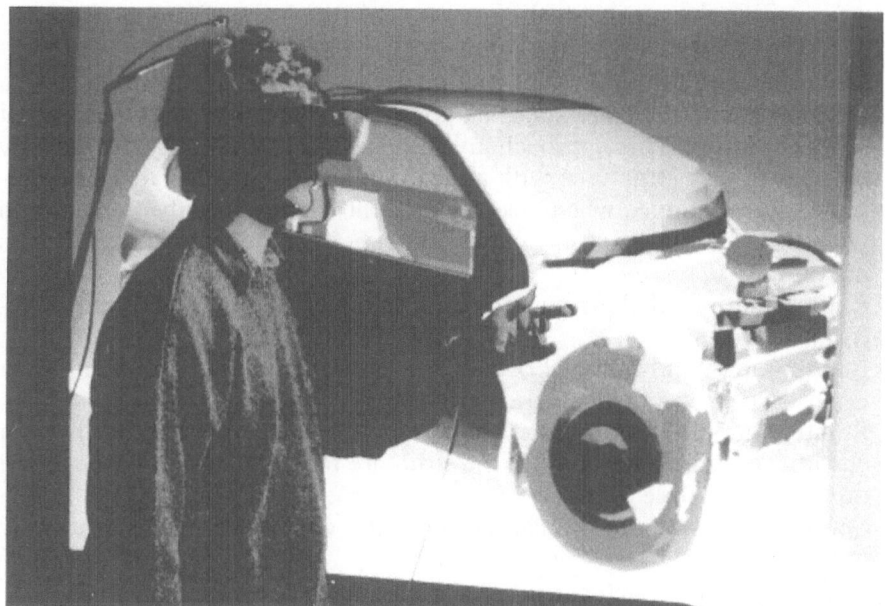

Figure 6.6. Visualization of crash calculations (FEM).

FEMWalk - A Dynamic postprocessor for the evaluation of results from FEM calculations

Crash calculations of a whole vehicle take numerous hours even on today's most modern and fastest computers and are therefore far away from real time.

For a better interpretation of crash calculation results and other FEM computations, we launched a tool - FEMWalk - which offers the user the possibility to move interactively around the vehicle (or other objects). In addition, the user can also observe the individual states of the computation (e.g. structural behaviour for a specific time) in "slow motion" by means of real time interpolation between the states.

The preparation of FEM computation results for the use in VR is quite easy. The standard form of so called "movie files" is supplied by all FEM packages. For a crash computation there are normally 25 or 30 states available (which means a state for every 2-3 ms for a total of 80 ms). In the preparation step the grid points of the structure for each of the states are converted and grouped into the VR format. This step normally takes no more than a ¼ of an hour.

Afterwards, the mechanical engineer and the designer can analyse the structure. FEMWalk offers the following possibilities :
- Interpolation of structural grid points between the individual states in order to create as many states as possible for soft and slow movement of the structure. This approach gives the user the possibility to view any number of part structures with almost any time frame in order to determine, for example, when one component touches another, when and where the points of stress occur and where the deformations form a maximum.
- Representation of deformations and tensions on deformed and undeformed shapes.
- Incorporation and removal of parts of the structure and groups in order to observe effects in a specific region.
- Relative and absolute movement of the observer; for example the user can attach himself to a part of the structure to evaluate the movement of other parts relative to his.
- Setting of time zones indicating the limits within which the results are to be shown.

Special situations, which can be prepared and later be called on again.

The virtual product clinic

In order to avoid faulty developments, vehicle concepts are tested within the framework of so called car clinics. This takes place long before their introduction into the market.

Up to now, such investigations were only carried out when expensive styling models created through many man hours were available. A short time ago, in a comparative test a virtual clinic was carried out by Volkswagen Marketing with the help of the VRLab.

The vehicle under investigation was the new SHARAN and the VR model was complete in interior and exterior. For obtaining a realistic impression, we used reflection mapping for the exterior as well as textures for the interior parts such as seats, instruments and floor.

The model was presented on a large stereo wall with a 3½ m diagonal screen and was moved by a 3D-mouse. In addition the doors and hatch could be opened and closed and the test persons could be seated into the driver's position.

The results generated by a typical marketing survey corresponded very well with the parallel tests on a real vehicle. Even the size relationships were evaluated correctly, although there was doubt that this question could be answered correctly because of the size of the stereo wall.

Of course there are many questions which cannot be resolved with a virtual clinic. For example questions concerning the smell, quality or surface structure (leather seats!). The seating comfort is another item which cannot be dealt with.

Figure 6.7. The virtual design of Sharan from Volkswagen (see section Color Plates).

On the other hand, however, the virtual clinic offers possibilities which the real product clinic of today cannot offer. Apart from the time advantage (long before the first real prototype is available) the cost advantage presents itself (one can envision, for example, a colour chart as well as a number of wheels which should be evaluated).

The manner in which design solutions can be influenced by the wishes of later customers is yet another advantage which already can be obtained at a point in time when development costs can still be kept in check.

6.2.4 Examples of future applications which are still in the development phase

Ergonomic Mock-Up

The ergonomic mock-up which is currently under development should serve for the individual evaluation of vehicle passenger compartments. Along with the evaluation of feelings of space, colours and materials it is also possible to investigate the placement and operation of instruments (for example navigation display position and ease of operation) as well as the view of the outside. In addition, the ergonomic mock-up is equipped with a steering wheel and all other necessary equipment, complete with force feedback, which is important for driving on an artificial road. The software for the driving simulation runs on a Pentium PC in parallel with the VR application.

Assembly/ disassembly simulation for manufacturing and service maintenance

Simulations for assembly and disassembly in the VR environment can be carried out in the field of manufacturing planning (ergonomic design, manufacturing sequences, feasibility, collision free design) as well as for service maintenance in the garage.

Prerequisites for these complex closed loop applications are precise tracking abilities and force- and tactile feedback. These are especially important in service maintenance simulations which can still not be verified successfully using today's VR equipment.

What can be done with today's possibilities are workcell simulations (e.g. with software supplied by Deneb Robotics or Technomatix ROB-CAD) as well as factory or manufacturing planning, which can be visualized excellently using VR.

A possible scenario for workcell simulation and visualisation:

A workcell for the spot-welding of a vehicle's body in white has been developed with ROBCAD. A number of robots are standing in the work chamber. Working together, the robots join parts and weld them. The spacial relationships in the workcell are to be determined if collisions between robots themselves or between robots and parts of the vehicle could take place.

The ROBCAD data (geometry, transformation matrices) are provided via an interface to VR (in fact, 2 different files). With the assistance of a suitable display like HMD, a stereo wall or a CAVE, and suitable interac-

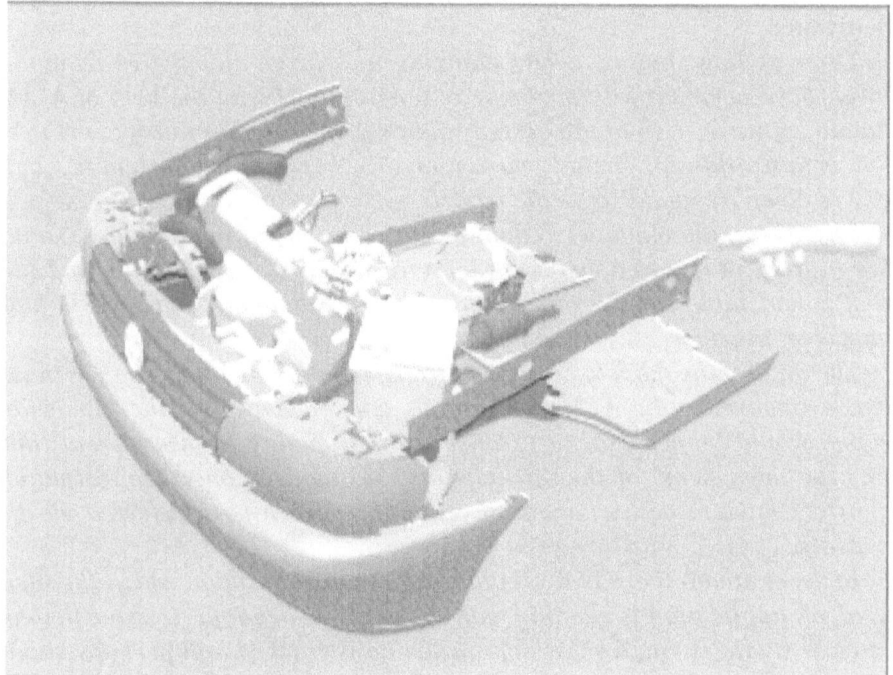

Figure 6.8. An assembly simulation scenario.

tion equipment like 3D mouse, data glove and speech recognition, the cybernaut moves freely through the workcell and observes the robots at work. In this walk, the collision status of previously chosen parts and work areas are constantly calculated in real time. If a collision occurs, it can be percieved optically and acoustically.

If it is necessary to observe a sequence in greater detail, the situation can be frozen. The cybernaut takes his virtual ruler and measures online the distance between 2 objects. The value is shown to him via his head-up display. For documentation purposes, all the movements and actions can be tracked, stored and displayed again in a later session once more.

Global Virtual Studio

The goal of a global virtual studio is to be able to gather in a virtual space around a model to discuss and make changes directly. This studio exists physically at several locations (e.g. design centres, studios of a big company). Only the movements and revisions are exchanged via fast networks using a suitable protocol, the visualisation is done locally.

A vision for a global studio which can already be realized with current hardware:

Three styling studios, each belonging to a large automotive manufacturer, have come up with a sample of a new model with the help of ALIAS Studio software. Two of the design studios are located in Europe, the third one is in California. The top managers, together with the designers would like to compare and discuss the models.

Physically, the managers could be found in one of the 3 studios standing in front of a stereo wall with integrated, half Virtual Reality and half real round table. In the other studios other managers and stylists have gathered, together.

The data of the models have been transferred to the other studios in advance since even the ATM connections cannot guarantee the bandwidth which would be necessary to transport high quality frames in real time. For the movements of the participants and models, only transformation matrices have to be exchanged. Eventually, speech is transferred via the network.

In every studio there is a graphics supercomputer upon which the identical VR application is running with the same model data. One participant in each studio is "active", which means he is tracked and perhaps shown as a dummy with live textures. The meeting starts, the active participant points at details, using a data glove or a virtual pointer. The other persons are looking through the eyes of the active person, as it were. Of course they can discuss and give their comments to each other.

6.3 Remarks on data related problems

6.3.1 Data exchange from VR to CAD

Up to now the possibilities of transferring data from a VR environment to the CAD world are quite poor. The main problem is that in VR the data are normally in the form of tesselated polygons whereas CAD data are normally in parametric form.

One day in the future, the CAD and VR worlds will grow together. However, as long as the worlds are still separated they must communicate via software interfaces. One direction, from CAD to VR, already works quite well. However, the fact must be noted that for the most part only geometric information is provided by the standard interfaces.

The other direction, from VR to CAD, is almost totally unavailable at the moment. One solution for the problem would be to generate parametric data in VR and tesselate "on the fly", in real time. This could be done successfully when parts or objects are generated in the VR environment (e.g. graphic primitives like boxes, cones, cylinders etc. or the center line of pipes).

6.3.2 Data interfaces

Today's widely used interfaces like IGES only provide geometric data. For example, they supply the surrounding of a hole but not the hole's information itself, the diameter and the position of the center. There is no doubt that for many applications we need considerably more information about materials, material properties, physical behaviour, colours and so on.

A good example for this lack of information is a fuel tank which has to be mounted into the body. As this tank is flexible, we need the physical data to make the decision if it could be assembled or not. Therefore, it is absolutely necessary for the future to work with product data models like STEP.

6.3.3 Data reduction

As long as the hardware is not powerful enough, CAD data not only have to be converted to polygons but have to be heavily reduced in order to get a sufficient frame rate. With intelligent reduction algorithms it is possible to achieve good quality with data reduction up to 10 % of the original supplied amount of data. Another frequently used method for data reduction is the generation of so called Levels-of-Detail (LOD's), which means the generation of different levels of complexity for one part. The LOD's are stored together in the data tree and are visualised depending on the distance to the viewer position.

6.3.4 Realtime versus complexity

Due to today's performance capability, the hardware must arrive at a suitable compromise between the realism of the depiction and the necessary image rate to permit interactivity in the VR environment. This of course is dependent upon the particular application and the immersion in the virtual world. Suitable texturing possibilities like reflection mapping help increasing the degree of realism without increase the number of triangles.

Literature

(Excerpt of recommended literature to become familiar with VR technology)

Kalawsky, R. S., The Science of Virtual Reality and Virtual Environments, Addison-Wesley Publishing Company, 1994

Burdea, G. C., Force and Touch Feedback for Virtual Reality, John Wiley & Sons, Inc., 1996

Computer Graphics Proceedings, SIGGRAPH Conference Proceedings, Annual Conference Series, Publication of ACM, 1992 - 1996

Computer Graphics forum, The International Journal of the Eurographics Association, Blackwell Publishers, Oxford, England, 1996

Presence, Teleoperators and Virtual Environments, MIT Press Journals, Cambridge, MA, USA, 1993 - 1996

7 3D Realtime Simulation and VR-Tools in the Manufacturing Industry

Dieter Bickel
Deneb Simulations-Software GmbH, Neuss

In the real world of Virtual Reality conferences participants often try to make a distinction between pure real-time simulation and Virtual Reality applications. The discussions on this topic often become emotional. The following quote of a definition of Virtual Reality might help one to understand that there is no sharp distinction between the two and both application areas need each other to successfully support the end users in the manufacturing industries.

Virtual Reality is an artificial environment, based on computer data, which accurately represents the physical aspects and the dynamic behavior of a model of the real world. Users and engineers are enabled to interact with the objects of the virtual environment using various virtual reality tools.

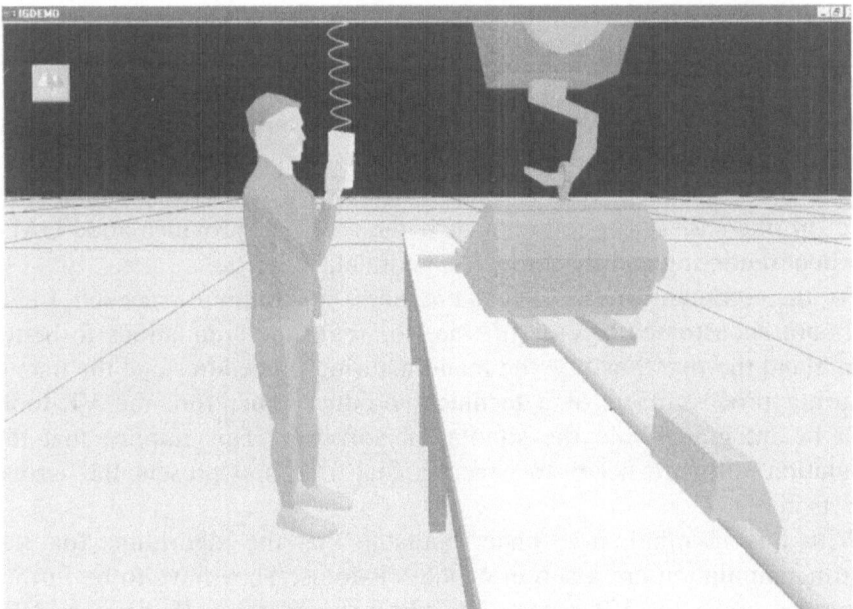

Figure 7.1. A virtual worker in a virtual production environment

Some well known VR-tools are the data helmet, the boom and the stereo shutter glasses for the visual aspects of the partial or full immersion into the virtual world. For grabing and touching, the data gloves are popular. They are often used as well to define directions for the fly through in the model. The placement and orientation of the real person inside the model is measured through tracking devices using magnetic fields or optical sensors.

In large research institutes and some very large companies the CAVEs have become popular, allowing the user to be in the middle of the virtual world and to share this mixture of real life and virtual model at the same time with other participants. While CAVEs need at least three stereo projectors for two walls and the floor, the Virtual Workbench only needs one, but still allows the user to walk around virtual objects and to look at them from different sides like hologram.

Traditionally, VR-tools were used in places where big budgets were available and profitability did not have the highest priority. These users came from the defense industries and related areas and not from the manufacturing industries. Now with PENTIUM PRO PCs and OPEN-GL graphic boards under Windows-NT, the hardware costs are decreasing to levels which even allow small and medium-sized companies to invest in VR hard- and software.

All the above mentioned tools do not expand the human audio-visual and tactile senses. There is no advantage over the method of looking to a real physical model of an engineer's design, beside the fact that the virtual model might be faster and cheaper to create. Nevertheless there are situations in architecture, car- and ship- design, where exactly these two facts are enough to justify investment into often expensive VR-hard- and software. In this case it is good enough if the software provides good CAD, excellent rendering and fly-through functionality.

For the communication between engineers much more is needed. Engineers are accustomed to use software for real-time simulations to better understand the functionality, the manufacturing procedures and the manufacturing process itself of a technical product. Therefore the VR-tools must be integrated into the simulation software. This implies that the simulation software heeds to peact in real-time and present the virtual objects in 3D.

With this in mind, it is understandable that the algorithms for 3D-realtime simulation are a subset of the VR-tools. They have to be further developed under the VR-aspects. Building up mental walls between VR-tools makes no sense.

Similar to other engineering fields, the main application areas for virtual reality tools including 3-D simulation software in the manufacturing industries are as follows:

- Simulation Based Design
- Virtual Prototyping
- Concurrent/Simultaneous Engineering
- Collaborative Engineering
- Planning of the Manufacturing Process and Procedures
- Training/Education
- Ergonomic Analysis/Design

In the following sections some examples are presented which should show how Deneb's simulation software is used by our customers in various application areas. Some of the examples shown are still from the defense market, but the reader should be able easily to apply them to other areas in which the resulting ideas of these examples could be of good use.

7.1 Simulation based design

Figure 7.2. Dynamic simulation of a landing gear

The above picture shows the landing gear of an airplane which is simulated in its dynamic behaviour with the Deneb software ENVISION. With this dynamic simulation the engineers can study the interaction between the various parts of the landing gear. Thus they can, for example, avoid collisions between these parts (Figure 7.2).

Figure 7.3. Loading of the LCAC-ship from water

The future user of a device or machine should be interviewed in a very early design phase in order to get his or her opinion and advise. This will increase the customer satisfaction later on. In the above example the end user could investigate in an artificial environment what the new product means and how it would perform in the real world. In this case it was the loading of the Navy's LCAC-ship directly from the water (Figure 7.3). Therefore the model not only needs 3D-geometry but also moving waves to more closely model reality.

The early investigation into system interdepencies is extremely helpful to avoid unnecessary changes in a later phase of the production, not to mention the high costs which are often inherent in these changes. In the above picture the upper part shows an overview of the situation at the quay while the truck is driving into the LCAC ship. The lower part shows the situation at exactly the same time, but from the view of the truck driver looking into the ship's body (Figure 7.4).

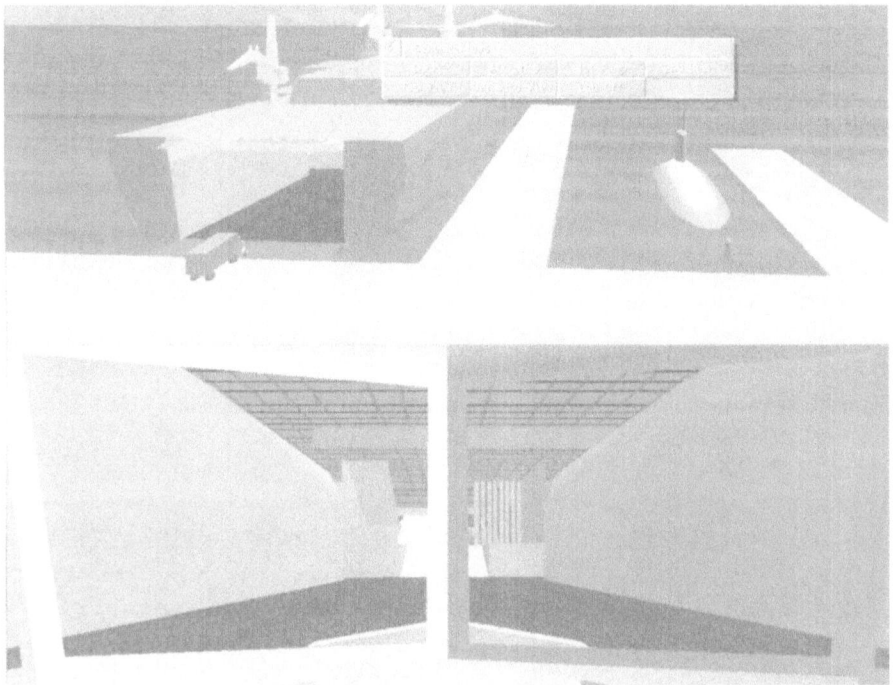

Figure 7.4. Driver´s view into the ship.

The Electric Boat Division of General Dynamics was able to reduce the design time by approximately 50% as compared to traditional methods, by using this types of development aid.

7.2 Virtual prototyping

Virtual prototypes allow alternative functional concepts to be used and tested.The costs involved are often essentially smaller than doing a similar test with real prototypes. The following picture explains this with examples of autonomous robots which should move around in a harsh environment (Figure 7.5).

Figure 7.5. An autonomous robot moving around in a harsh environment.

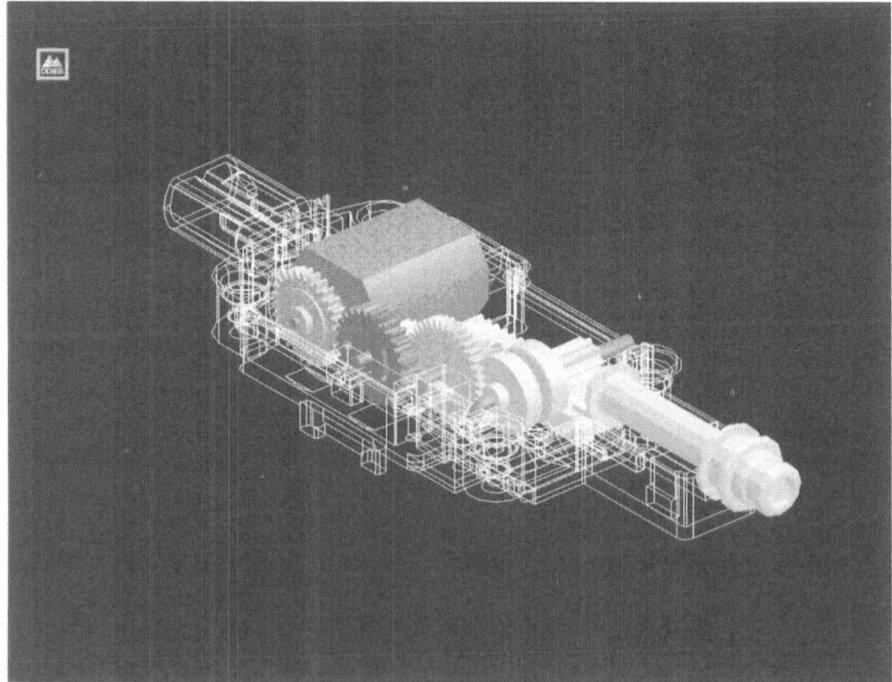

Figure 7.6. Inside of a car door lock.

A car door lock was developed as a virtual prototype by General Motors using Deneb's software. The 3D visualization of the movements between the individual parts of the door lock allowed for very early discovery of functional deficiencies. Those were caused by falsely positioned gears. This early discovery saved millions of dollars in warranty repaires which might have happened after the cars were sold (Figure 7.6).

This example leads to the next topic, since the simulation of the movements made the early discovery of field problems possible.

7.3 Concurrent engineering

Figure 7.7. Simulation of a robot workcell.

Concurrent Engineering supports companies to test and evaluate production alternatives before the design of those parts or products is terminated. In the factory Wörth of Daimler Benz since the Summer of 1992 engineers have used simulations of the workcells to decide if the welding points specified by the designers could be reached and manufactured with the KUKA robots in place (Figure 7.7). If not, welding and fixtures have to be

redesigned. More or less from the beginning the users in Wörth are applying stereo shutter glasses for better understanding of the complicated
situation between welding and truck cabins. Since the Summer of 1994
they have been doing all spotwelding programs off-line using Deneb's
IGRIP and Ultra-Spot.

This example shows as well that it is not sufficient to rely alone on the
feasability studies and the training of the future operators to justify such
an investment in hard- and software. The process know how itself and its
translation into productivity tools is an essential additional need.

In the case of the factory in Wörth the spotwelding programs for around
400 different truck cabins had to be generated in the first years of that new
factory concept. The robot operators estimated an effort of around 20,000
hours of robot programming online. In the beginning they taught one side
of the cabin. They uploaded those programs into Deneb's simulation, and
the programming system IGRIP mirrored them off-line, mading changes
and additions off-line as well. Doing this, the users in Wörth learned after
a view months that they could save at least 10,000 of the estimated 20,000
hours. In the end they saved even more. In the meantime they feel safe
doing all the programming off-line.

This off-line programming and optimization of the robot's work reduces production interrupts and allows for better usage and workloading
of the individual robot in the workcell. One reason for this better usage of
the robots and reduced cycle times for the whole workcell is the fact that
in the simulation it is possible to run the virtual robots with a higher level
of risk and in shorter distances than an operator would dare to do in real
life.

This precision of the off-line programming is only possible because the
simulation system and the robot controllers use the same path planning
algorithms. This is possible because the European car manufacturers
started the RRS initiative in the beginning of the 90ies.

7.4 Collaborative engineering

In end user products more and more components are produced by different
subsuppliers. During the design process engineers of the different companies involved should have the opportunity to work in different locations
on the same digital model of a product at the same time and to communicate on that work. Many simulation systems in the market allow this by
only sending the changes of the digital model and its view points between

the different computers. The visualization of the model and its view points is then done on each linked computer simultaneously, thus giving the users almost the impression that they are working together on the same computer.

Many public presentations are currently sponsored by public research money. Deneb's software is capable of this as well. The reality is still unfortunately not as advanced. One of the reasons for this might be the fact that it needs a lot of continuous work on the side of the system operators to keep up with changes in the application software and the operating systems. In the reality of Virtual Reality the compatibility of hard- or software components is still a challenge. This is even more extreme in heterogenions network environments and between different company cultures. Nevertheless technically it is state of the art.

7.5 Planning of the manufacturing process (Process Know-How)

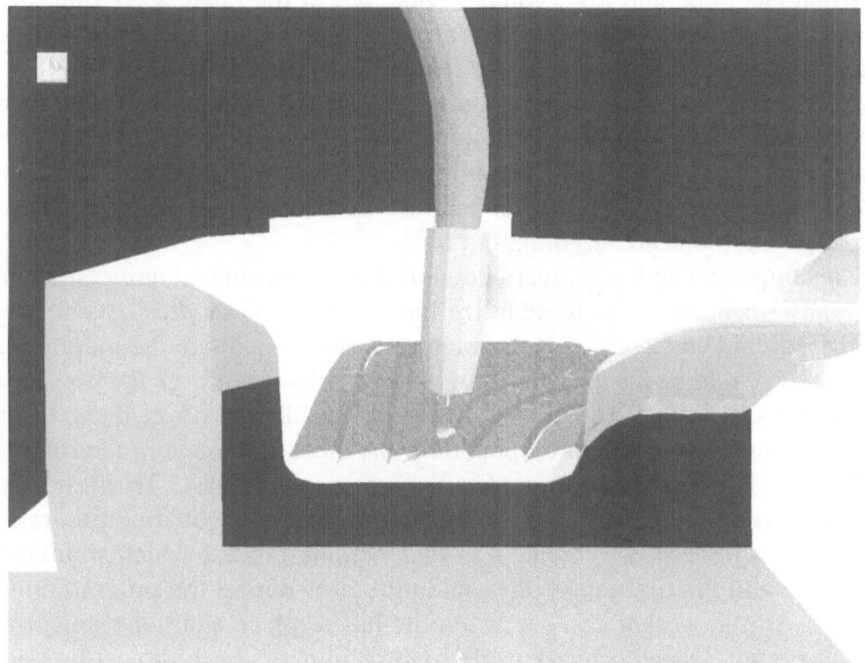

Figure 7.8. Simulation of die-reforging (1)

The above picture (Figure 7.8) is a cross-sectional snapshot depicting the simulation of a robot program, specifically a REIS robot for die-reforging. This program is totally done off-line. In the cross section one can see a black area and a gray area of the die. The picture also shows the welding pistol and the simulated surfaces off the welding seams which are placed by the robot. What is the task for this robot program and what is the process know-how behind it?

The die is used as one half of a negative form of an iron part which is pressed and thus formed between these dies. Those parts are subsets of the car's suspension and therefore have to carry very high loads and should not break. If so, the driver could, in extreme situations, lose control. In our case the user of the software is the company *cdp* in Daun, Germany, a subsupplier to BMW, Ford and others.

The gray part represents the „normal steel" volume of the die, while the black part represents one which consists of a specialized, hard steel, and is capable to resisting the extreme forces and temperatures during the pressing process. Until 1995 those dies were produced in the following way. A skilled worker was smelted the black material out of the gray block using a carbon electrode and high voltage. This part of the process continues to be done even today. After that a skilled welding expert was filled up the black volume with single welding seams. This stage is more or less finished when the negative form of the target part is reached. The die is then used between 8,000 to 30,000 times, depending on the type of part produced and the quality of the welding work. This work is done while the block of the die is kept at a temperature of 500 degrees Celsius. These working conditions are anything but nice.

The company cdp has cooperated with different technical universities in Germany for many years to stabilize the quality of these dies. Eventually the Technical University of Kaiserslautern tried to stabilize the number of times a die could be used by teaching a REIS robot to reforge the die. This teaching took around 160 hours, while the welding process itself takes about 8 hours. This was unacceptable for the industrial usage of the robot.

The need for off-line programming was growing fast. To allow the skilled workers to use their experience while doing the off-line programming, it was necessary to create a virtual welding process which simulates the shape and distribution of the welding seams during the programming process. So the worker can see where he has to place additional material and where not. Deneb offered a solution in a joint research project in 1993 which generated the robot programs interactively in about 80 hours. Al-

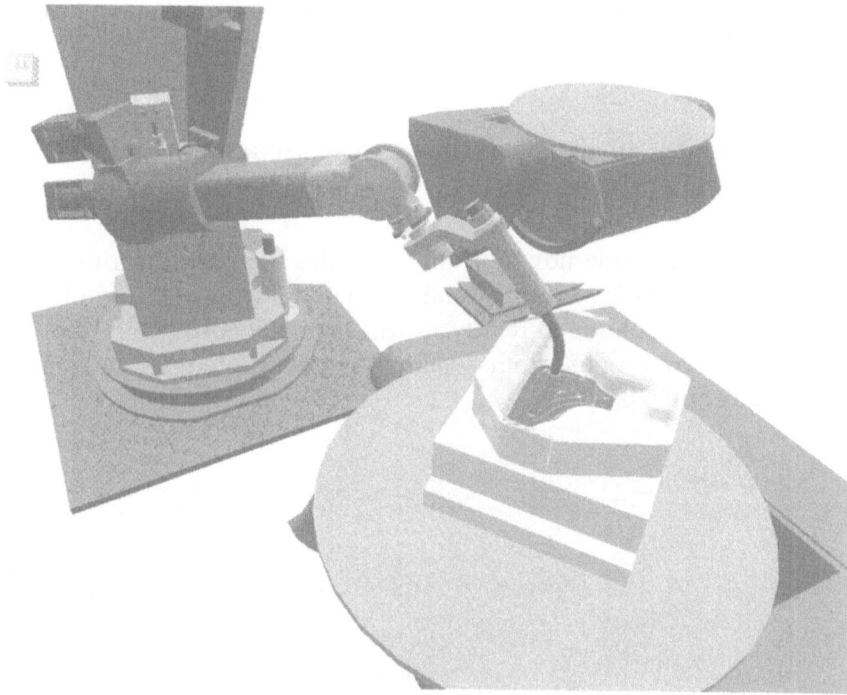

Figure 7.9. Simulation of die-reforging (2)

though this was a big step ahead and the robot was not unproductive during the programming, it was still not yet practical enough.

The company cdp, a privately owned SME, took the risk and cooperated further with Deneb to develop a more automatic generation of the robot paths. Deneb could solve this task in a reasonable time frame by using its macro language capabilities and system openness. The macros were developed through intensive interviews with the supervisors of the welding staff of cdp and also with the experts of the robot producer REIS.

Today the official time frame to program such a die is around 16 hours, while the welding process itself is also reduced to about 4 hours. All programs are only done off-line.

Deneb and cdp are still attempting to come closer to the reality with their virtual welding cell. The bottleneck here is the lack of theoretical algorithms to calculate the precise flow and shape of a welding seam in a 500 degrees Celsius environment during the welding process. Here new physical algorithms of the real-time simulation could improve the closer-to-reality aspect of the Virtual Reality. This is also a typical example of how simulation techniques can help to improve the productivity of a company and how the quality of the products can be increased.

7.6 Planning of the manufacturing procedures (Factory Simulation)

The production know how not only includes the process in the workcell but also the working sequences and the material flow of the whole factory. To improve the closer-to-reality aspect of a Virtual Factory the need to incorporate algorithms from Operations Research into the simulation software is obvious.

Figure 7.10. Virtual factory simulation (see section Color Plates).

The Virtual Factory simulations are often used in a very early phase of the layout design of a factory. This should happen before the architectural drawings are produced. That could be another example as well for concurrent and collaborative engineering. As architectural work, this type of 3D-simulation is often done as engineering and consulting from smaller engineering firms. Bigger companies like the car manufacturers have this type of engineering in-house. In 1997 the market in Germany has shown an increasing interest in the combination of discrete event simulation and 3D-objects using different levels of immersion. One reason for this is the fact that software from Deneb for this application is available not only on expensive workstations but also on PCs under Windows-NT. More and more small engineering firms can affort to buy it.

In the Summer of 1994 AUDI purchased engineering from Deneb to simulate the whole plant in Ingolstadt. The purchase order was based on 26 man-days from Deneb and AUDI added a similar number of man-days to it. The hardware consisted of an Indigo 2 Extreme which had a performance of a DM 12000,- PC of 1997. Consequently the geometric modeling was rough, because the simulation had to move thousands of cars through the model.

Even with this simplified 3D-geometry the understanding of the different situations in the factory was very good and the communication between the different planning experts was shortened and improved a great deal. Most topics of the discussions between the experts could be animated and simulated on the screen of the workstation. Following the statements of AUDI, this simulation saved millions of DM.

7.7 Training and education

Training and Education are natural applications of Virtual Reality. In this field reality would be the best teacher, but in reality errors can be deadly. In Virtual Reality no one can really die.

The profit of the Virtual Reality usage in training and education is even more difficult to measure than in other fields. Therefore, considering toady's equipment costs of Virtual Reality in the manufacturing industry, it will only be used as an add-on to its usage in other areas. One of these main areas of usage might be advertising.

Again AUDI in Ingolstadt is an example for this mixed use. In December 1996 the foundation for a new painting plant in Ingolstadt was laid. The company saw the need to show a video of the production of this future plant. Deneb was contacted in November 1996 because of its 1994 simulation. Deneb offered to make a digital model of the whole plant in about 9 man-week. The simulation was based on 2D-DXF files, drawings from the architects and 3D-models of the painting machines from DÜRR in the ROBFACE format, which Deneb's 3D discrete event simulation software QUEST could read.

To assure that the assembly lines produced with the right speeds and numbers, in addition to the 3D modeling the logical modeling of the plant also required an essential part of the 9 man-weeks. But the video was ready just in time and the discrete event model lives in various computer types. The same day of the presentation of the video to the public, an essential portion of the AUDI management was trained with the 3D discrete event simulation in sessions of 30 minutes with partial immersion using a stereo projector and shutter glasses.

Meanwhile AUDI purchased the software, the digital model and a high end PC to demonstrate the painting plant to visitors in their visitor center. This is a good example of mixed usage for advertising and training.

Based on the fact that prices for hardware are still falling around 40% per year, it can be expected that Virtual Reality applications will be used in

Figure 7.11. A fire fighting training scenario (see section Color Plates).

training and education even in smaller companies in the manufacturing industry. The interesting examples today come from the defense industry.

The above picture (Figure 7.11) shows a fire fighting training in a submarine using Deneb's software.

7.8 Ergonomic analysis/design

Here the simulation and off-line programming of robots deliver the master copy from which feasability studies, full posture analysis and cycle time calculations for human work are derived. In addition, appropriate software algorithms generate figures on lifting limits and absorbed calories. Under- or overload of muscles might result in permanent injury of the person performing this task. For this reason detailed posture analysis is necessary to keep a company's manpower healthy. Deneb's software gives the engineers algorithms for revised NIOSH lifting, Garg's energy expenditure prediction model and RULA posture analysis. Thus together with the 3D-simulation, Deneb's ERGO package is going far beyond the limits of normal timestudy software.

Figure 7.12. Ergonomic analysis.

With the close-to-reality simulation capabilities of Deneb's human models it is possible to design, among other things virtual machinery for the assembling of parts above the head of the worker. Concurrent engineering and all the other topics introduced at the beginning of this article are back on stage.

7.9 Conclusions

The application fields of Virtual Reality mentioned in this paper support and overlay each other. Over time they will enforce strong organizational changes inside those companies which make use of them.

In Germany by end of 1996, only 10% of all CAD-systems use 3D geometric entities. That is the real bottleneck for the introduction of Virtual Reality into the manufacturing industry. Let's concentrate on increasing that, to pave the road for Virtual Reality.

Conference proceedings

Brown, R. G., Deneb Inc., An Overview of Virtual Manufacturing Technology, AGARD conference, May 1996, Sesimbra, Portugal

Scotton, T.W., United Technologies Research Center, Virtual Manufacturing, Deneb User Group 1995 Proceedings, October 1995

Annelle, J. and Braun, R., Grumman Aerospace Corp., Graphic Simulation System for Product Process Development, Deneb User Group 1994 Proceedings, October 1994

Bickel, D., Die Virtuelle Fabrik: 3-D-Objekte als Basis einer realistischen Fabriksimulation im Zusammenhang mit Top-Down Entwurfsstrategien, AWF-Kongreß, Germany, November 1994

Technical papers

Slawski, D. and Boudreaux. J., Full Speed Ahead, Virtual Environments and 3-D simulated models buoy development of next-generation submarine, CiME, Spring 1996

Beaudreaux, J., General Dynamics-Electric Boat, ARPA SBD Contract

Demmy, G., Navistar International Corp., Navistar Leads the Way with Virtual Manufacturing

McDonald, M., sandia National Laboratories, Virtual Collaborative Engineering, 1996

Jungfleisch, J. and Weiler, B., cdp Peddinghaus, Einsatz der graphischen Simulation beim automatisierten Auftragschweißen, Daun, Germany 1995

8 Virtual Reality in Telerobotics Applications at CAE Elektronik GmbH

Hans Josef Classen
CAE Elektronik GmbH

Space qualified robot control systems have to provide a broad range of handling capabilities and meet the demand for highly autonomous operation. The complex operating environment requires a sophisticated user interface whose primary task is to provide the user with complete control over the robots after a short learning phase. These requirements cannot be fulfilled by interaction media like a keyboard, a joystick, and a spaceball.

In 1991 the idea was born to study the benefits of Virtual Reality (VR) as user interface for teleoperation of a robot system in space. At that time VR was promoted as a new technology but, actually, had already been used in CAE's flight simulators for decades.

The project *VITAL*[1] was set up as a research study within which the usability of VR in teleoperation was to be investigated. As the direct successor project in 1994, *VITAL-II*[2] had the goal to clear the drawbacks that were detected in *VITAL* and to develop a VR system for external servicing of satellites. In the project *MARCO-X*[3] the VR technology developed in *VITAL-II* is used to control a robot system different from the *VITAL-II* robot system.

This paper describes the efforts that have been undertaken at CAE Elektronik GmbH within the above mentioned projects. In addition, two more projects are outlined in which the CAE VR technology has already been applied.

Keywords: Telerobotics, Virtual Reality, Virtual Environment, Remote Operation

---

[1] The project VITAL was supported by the German Space Agency (DARA) under contract number 50IP9176 and the State of Nordrhein-Westfalen (NRW) under contract number I/C 322-9111a46

[2] The project VITAL-II is funded by the German Space Agency (DARA) under contract number 50TT9434-AK

[3] The project MARCO-X is funded by the German Space Agency (DARA) under contract number 50TT9507.

8.1 Introduction

VR as a paradigm for man-machine interface to telerobotics has been used at CAE Elektronik GmbH since 1991. Under contract by the German Space Agency DARA three projects were performed at CAE Elektronik GmbH.

The first project, VITAL, was targeted in the investigation of the usability of VR in the area of telerobotics and telepresence. A state-of-the art VR system was used as a base to control a robot system made up of two robots in a re-created space testbed. Results of the research showed that Virtual Reality was generally useful for teleoperation, and hints were produced where further developments were necessary.

Based on this knowledge the development of a VR system that is capable of working in a partially unknown environment was started in the

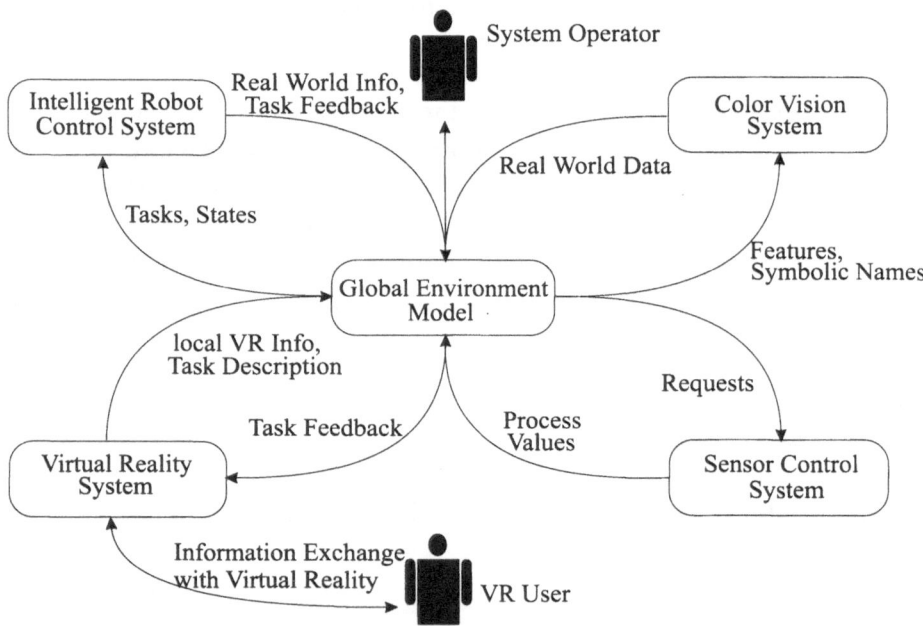

Figure 8.1. Design of the VITAL system

project VITAL-II. This system allows the online adaptation of the environments, the robot control system, as well as the VR system to reality by the use of a laser ranger and a vision system. Due to the adaptability of the environments, the system can react to problems which are not pre-coded into the database. Furthermore, the VITAL-II VR system is designed to be able to control different robot systems which appear to the VR system kernel as front-ends.

The VR system which is going to be developed within VITAL-II is used in the parallel project MARCO-X in that the robot of the ROTEX space experiment will be controlled by CAE Elektronik GmbH's VR system. The goal of this project is to harmonize the developments in the DARA projects MARCO and VITAL-II.

8.2 VITAL – the application of Virtual Reality in telerobotics

8.2.1 Goals

The main task of the project was to improve the man-machine-interface between a mission specialist on earth and a robot laboratory in space by means of VR. The mission specialist should be able to perform his experiments in a virtual laboratory in an intuitive manner. His inputs are transmitted to the laboratory in space and are carried out by the robot system. The key idea is to de-couple the scientist from the robots: the mission specialist can concentrate on his experiments and must not care about robot kinematics or interaction media like keyboards or joysticks. The VITAL system was designed around an off-the-shelf VR system manufactured by VPL. The reason for this was the fact that within VITAL the usability of VR should be investigated and no VR system should be built. VITAL took place from September 1991 until August 1994.

8.2.2 VITAL system design

The center of the VITAL system was the Global Environment Model (GEM, Figure 8.1) which held all data in the system. The user communicates with the VR system, commands derived from his actions are sent to the robot control system which constitutes the transformation to the real world. The Intelligent Robot Control System at the University of Dortmund that controls and coordinates two robots, was connected to the VR. A color image processing system was designed to monitor the situation in the laboratory, for instance, to ascertain whether a door that should be

Figure 8.2. VITAL Virtual Laboratory (see section Color Plates).

closed by the robot is actually open. A sensor control system was used for measuring and actuating the controls.

The VR system was equipped with an LCD-based head mounted display and a dataglove. The dataglove was extended in the course of the project by intergrating tactile feedback into the VPL dataglove. The glove uses 14 air cushions. Each cushion is pressurized when the user touches a virtual object with the respective part of the finger and/or hand.

Stereo images were generated by two Silicon Graphics 4D/310VGXT's interconnected by Ethernet. The system was controlled by an Apple Macintosh. In addition to the dataglove a spaceball could be used to control the user position.

The STATEX-II experiment of the German D2 Space Shuttle mission was chosen to demonstrate the capabilities of the system. The experiment was reduced to the main handling steps that had to be performed, on the one hand, by the mission specialist in the virtual laboratory and, on the other hand, by the robot system in the laboratory.

Figure 8.2 shows the virtual scenario of VITAL. It is made up of a copy of the robot laboratory with its doors, levers, plugs, drawers, and the experiment container itself, but without the robots. The mission specialist was enabled to move through the virtual world with the help of the spaceball or hand gestures. Special gestures were defined to manage the robot lab (open doors, close drawers, etc.) and to handle the experiment (move the experiment container, inject fluid, connect plugs, etc.). Tasks for the robot system were derived from the actions of the user in the virtual world and sent to the robot system.

8.2.3 VITAL results

The main result of VITAL was that the robot control benefits from the use of VR as ist user interface. VR provides an intuitive shell around complex control structures. Interactions are initiated by the user as in the real world, feedback from the system is presented to her/him at the same level. The virtual environment as the user interface can be generated to be convenient and easily comprehensible for the user and does not necessarily have to be modeled exactly like the physical environment, which was built to „accommodate" the robots.

The „virtual" STATEX II experiment was performed by many people with different technical backgrounds who stepped into the virtual world and were able to work only with the help of a storyboard. It showed that the air cushions in the feedback glove were very important and made the virtual actions safer and faster. The spaceball extended the system in so far that user movement was done using the spaceball with one hand, interaction was done using the dataglove with the other hand. The combination of the spaceball and the tactile feedback glove as interaction media with the virtual world enabled the user to perform the virtual experiment that was carried out by the robots in response to commands coming from a virtual world.

The implementation of the developed concepts into the VITAL system was very complex. A number of very heterogeneous systems had to be integrated into one functional unit. A dedicated communication protocol was developed which copes with the control and feedback problems on an abstract level. The environment description on a separate system provided for up-to-date system states. The functional VITAL system was successfully demonstrated in August 1994.

In summary, the experiment was fully coded into virtual reality. The robots could be controlled by an intuitive man-machine-interface. Although the off-the-shelf VR system structure was not very open and the system had limited capabilities in some instances, it could be fully integrated into the complete VITAL system.

Using the communication protocol, all systems could be addressed quickly, safely and precisely. Another advantage was the use of identifiers for virtual and real objects, allowing a unique identification scheme on the one hand and (perhaps) totally different naming conventions in the subsystems. A global environment model with its data and control flow monitoring served for the correlation of the different world models in the subsystems. The system structure had been defined in order to allow several or different systems to be controlled very easily.

The major limitation of VITAL was the fact that the robot scenario had to be known in advance. In an unknown or only partially known environment, the described approach was not feasible. VITAL provided teleoperation (coupled with the difficulties due to different kinematics in man and machine and the transmission delay between earth and space) to cope with problems of uncertainty and error correction, but with the teleoperation-approach the intuitive user interface was lost.

8.3 VITAL-II

8.3.1 Goals

The prime target of VITAL-II is the ability to cope with partially unknown environments which appear, for example, during external servicing of satellites. A computer vision system provides for a semi-automatic update of the virtual environment of the VR system as well as the robot control system. Thus, the robot control system and the VR system had to be extended to allow the environment update.

In the case of the VR system this meant a completely new design since the one used in the precursory project VITAL did not provide for this. Consequently, a design goal was an open system easy to port to different hardware and easy with which to interface. It is now based on the SGI Performer library and makes heavy use of structuring the virtual world. A key enhancement is the possibility to update the environment, which is a unique feature of a VR system. During the update new objects can be added (including links to parent and/or child objects) and objects can be deleted. In addition, position and orientation of objects can be changed which, of course, is normal in VR systems/libraries. The newly generated objects can be divided into 2 groups: simple objects and deposits. With this distinction new commands for the robots can be generated without having to change the processing of the robot control system.

In addition, the robot system is being expanded to control redundant kinematics.

The project was started in August 1994 and is due to end in February 1997.

8.3.2 VITAL-II system design

The VITAL-II system is designed around a Central World Model (CWM, Figure 8.3) which assures consistency and correlation of global data in the complete system. The CWM is built on top of a relational database management system which is enhanced by functions not yet available in standard SQL: e.g. trigger functions and message passing functions. All subsystems, the VR system, the Vision System, and the Intelligent Robot Control System are connected to the CWM and exchange information and commands only through the CWM. Contrary to the GEM of VITAL the CWM is an active part of the system and is able to trigger other systems if data have been changed. This was not possible in VITAL, where the GEM

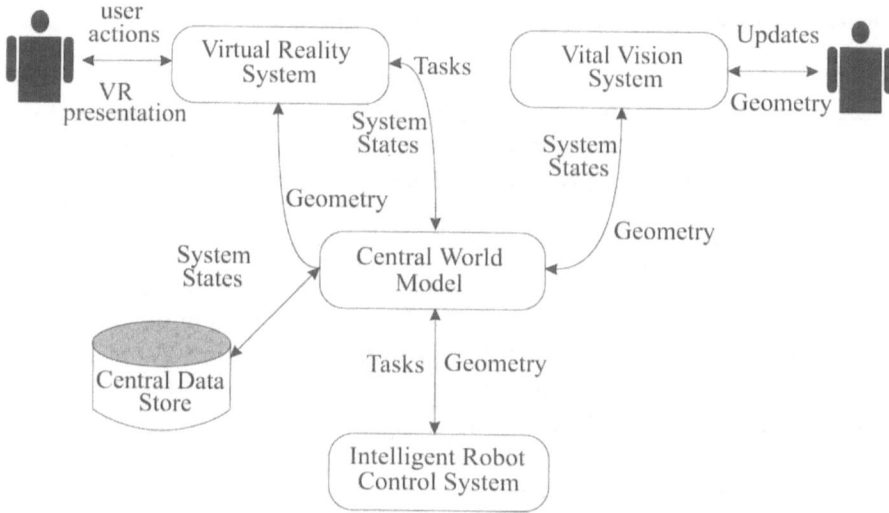

Figure 8.3. Design of the VITAL-II system

was only used as a data store.

The CWM contains scenario descriptions for the robots and for the VR and the states, positions, orientations, etc., of all real and virtual objects. Triggers can be defined that notify a subsystem of changes in the CWM data. E.g. if the robots have changed the orientation of a flap then the VR system is notified in order to perform the change in the virtual environment, too.

Figure 8.4. The VITAL-II virtual laboratory with the satellite testbed

The VRS receives the scenario geometry from the CWM and uses it for the generation of the virtual environment (Figure 8.4). Reconstructed surfaces are loaded from the CWM as well. The VRS decodes user interactions and issues commands to the robots via the CWM. Feedback is displayed respectively.

The Vision System employs the robot control to move the cameras and the laser ranger. A scenario update is reconstructed interactively. The first step is the reconstruction of a surface which is reworked in the VR system by adding or deleting faces or other primitives. The final result serves to update the operator's environment.

Other improvements include the integration of the dataglove control with the generation of feedback. Both systems are now controlled via a common bi-directional interface.

8.3.3 VITAL-II evaluation

Due to the ongoing effort in VITAL-II it is too early to give an exhaustive evaluation. The major drawback identified in the precursory project VITAL, however, has been solved: the scenario can be updated according to reality. Teleoperation is no longer necessary to try to fix any unforeseen difficulties. All robot actions can be handled in the preferred Automatic Mode, which can be completely controlled by the VR system.

The following scenario is possible:

A satellite with improperly clamped solar arrays hads to be repaired. The basic structures of the satellite are known; it is not known, however where the problem, i.e. a clamp, lies. With the help of the vision system it is possible to detect the clamps, update the environments of the CAE VR system and the robot system. The update includes a new object „clamp" which can be extracted in a certain direction, and the solar array can be extended. In previous VR system one would have had to switch to teleoperation, which does not make life easier.

The VITAL-II VR system's second important feature is that it is now running under a standard graphics library and can easily be used on workstations with different equipment. Even porting it to more advanced software and/or hardware equipment is quite simple. Furthermore, a number of data format import filters are available which assure, for instance, data import from modellers like MultiGen.

In its current version the VR system uses an open design which allows for the easy integration into a broad range of applications. It serves as the base software for CAE VR applications.

Figure 8.5. ROTEX testbed

8.4 MARCO-X

The first application of the CAE VR system was the control of the MARCO robotic subsystem. This component is the ROTEX robotic controller from the German DLR institute which flew in the German D2 Space Shuttle mission.

In order to connect the VR system to the controller, work had to be performed in two areas: modelling of the scenario and commanding the robot.

Figure 8.5 shows the MARCO-X scenario with an Orbit Replaceable Unit (ORU) and three TRUSS structures. It was imported into the VR system with the help of two versions of the Inventor file format import filter. In contrast to VITAL-II, TRUSS structures can form a „tower"; speaking in the VITAL-II language: they can be moveable objects and deposits at the same time.

Command generation is totally distinct from the commands on an abstract layer in VITAL-II. In addition, the generated commands in VITAL-II are written in ASCII, where as the robotic controller accepts any binary codes.

Even though the command semantics are much different from VITAL-II the additions to the CAE VR system were minor. One can switch between VITAL-II and MARCO-X on the user interface of the VR control workstation. The renderers switch between the environments and command the robots accordingly.

The final presentation will be done via an ISDN connection between CAE in Stolberg, which is near to Aachen and, the DLR in Oberpfaffenhofen. It is scheduled around the time of this workshop.

8.5 Related applications

The CAE VR system has been used in various projects.

The first application was the visualisation of a rendezvous and docking operation between a chaser and a target satellite in the DARA project VENUS. A simple environment without any connection to a robot system was created and animated through the help of a simulation. The user was able to look at the scenario from arbitrary positions.

Another application was the use of the VR system as a design aid for an environmentally friendly aircraft washing system. Figure 8.6 shows the virtual scenario that was used in the *WARO* project.

Figure 8.6. WARO virtual scenario (see section Color Plates).

The simulation of the four robots was performed in RobCAD and was imported into the CAE VR system via text files. The same holds true for the visual scenario. The aircraft was modelled according to drawings.

Equipped with a head mounted display the user is able to walk through the virtual world and is able to judge if the brushes at the end of the robot end effectors really touch the plane. The path is somewhat predefined as in real scenarios in order to prevent the viewer from crashing into a robot or the plane.

8.6 Conclusion

Being the latest VR project at CAE, VITAL-II extends the work towards an intuitive user interface for teleoperation of robots, especially in space. The VITAL-II system provides unique functions for the interactive update of scenarios that cannot be found in other operational systems. With the VR system, the first steps towards a 3D modelling software are under development. The Vision System provides the combination of presentative graphics and generative graphics. The easy applicability of the concepts to different tasks and/or environments has been proven in various projects.

Last but not least the robot control system provides key technology to support the interaction between physical and virtual environments with the help of robots. Thus it might be opening the door for several new applications in fields where traditional „teleoperation" is used today, and for new fields like e.g. in medicine.

The application of the CAE VR system in the project MARCO-X on the one hand shows the adaptability of the system to different operational requirements and on the other hand proves the concept.

The project VITAL-II is handled under ESA PSS-05 guidelines. It is currently in the integration test phase. The first prototypes are running, integration tests with all subsystems will be performed via Internet within the next months. Final presentation is scheduled for April 1997.

9 Experiences with Virtual Reality Techniques in the Prototyping Process at BMW

Antonino Gomes de Sá, Peter Baacke
BMW AG, Munich, Germany

Physical prototypes or mock-ups are a very important evaluation instrument during the complete design process. In the automotive and aeronautical industries, CA technologies, have been used more and more during recent years. Due to the fact that the setting up of a physical mock-up takes from 8 to 12 weeks, it is possible that in this period there will be many changes in the designers' CAD models. The physical mock-up represents an old design stage as soon as it is set up.

Today's strategy is to replace a high proportion of physical mock-ups with digital models - the so-called Digital Mock-Up (DMU).

To support the master build today, conventional CA and simulation technologies (e.g. CATIA, ROBCAD, ViW[1]) are intensively used in the prototype vehicle group and package department. Thanks to installation and assembly/disassembly simulation which take both manufacturing and service aspects into account, the expenses of physical mock-ups can be reduced dramatically.

Complete process assurance is not possible with conventional CA technologies for the following reasons:

- No intuitive handling of components and tools
- No 3D impression at more complex installation situations
- Poor interaction techniques
- No reality-relevant response times for interactions (e. g. collision identification)
- No simulation of flexible components (e. g. door weather strips, wire cables).

Due to these facts, Virtual Reality (VR) technologies have to be developed to increase the areas of interest and to set up these technologies as a key building block of the Digital Mock-Up strategy.

---

[1] ViW - Virtuelle Werkstatt (Virtual Workshop Software), a trade mark of Tecoplan Informatik GmbH, Munich.

9.1 Project definition

In September 1995 a project for testing VR technologies to master-build assembly studies was initiated within the framework of the ESPRIT Project 7704 AIT - Advanced Information Technologies in Design and Manufacturing[2]. The project was organised by BMW and Rover representatives and Prosolvia Clarus AB, a Swedish VR development company [1].

On the basis of two application scenarios - installation of the door lock and the window regulator - the feasibility of installing these components into the door (Figure 9.1) was examined. The door lock scenario is representative of today's fitting and build-in simulation in the prototype vehicle group. It contains many complex geometrical parts, complex fitting paths and difficult accessibility of components, tools and the worker's hands.

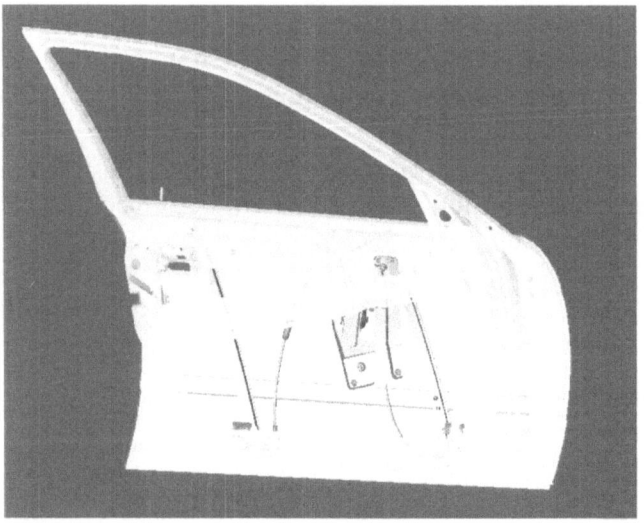

Figure 9.1. Door with installed lock and window regulator (see section Color Plates).

9.1.1 Project targets

The objectives of the pilot project were to test a VR system, consisting of real assembly and disassembly processes, to show the VR state-of-the-art

---

[2] AIT Project (ESPRIT Project). The project partners are: Aerospatiale, BMW, Saab, BMW Rolls-Royce, Audi, Alenia, Dassault Aviation, Britisch Aerospace, Renault, Reydel, Mercedes-Benz, CASA, PSA, VW, Magneti Marelli, Fiat and Rover

and the limits of today's graphic machines and to evaluate whether VR-technologies are the right IT tool to fulfil the user requirements in the prototype vehicle group. Another important aspect was the applicability of VR as a decision-making platform.

9.1.2 Implementation

The models used were originally designed in CATIA (Dassault Systems, France). The models are created with surfaces and solids. The topological elements are described as NURBS - a mathematical surface description. This is not a valid representation for real time graphics, since it involves solving the equations in order to render the surface. The model needs instead to be represented by a fixed set of triangles, which then can be sent to the rendering pipeline. The number of triangles sent to the graphic pipe will affect the rendering time, and hence the frame rate. [2]

In order to convert the NURBS models from CATIA to polygons, the models with surfaces were first exported to the IGES format, which retains the NURBS information. These IGES files were then read into the Clarus CAD Real Time Link [3] [4]. Here these surfaces were interactively tessellated using different triangulation algorithms, levels-of-details (LOD) could be created to increase the performance or set various colours and textures.

In this project IGES data were used for all surface parts. For the solid models, such as a door lock or window regulator, ViW Software[3] was used. This format is used by the software of the Tecoplan Informatik GmbH company in Munich [6]. It is a polygon format, which means that this model will serve as the original for the reduced models. The highest level of detail can thus never contain more information than the voxel model.

The way the virtual world responds to input from the user has to be defined separately, in the VR system VEGA [4][5]. What should happen if two objects collide, or if one object is released in mid-air? Should it be affected by gravity? Each object, e.g. the door lock and the window regulator, can be manipulated by the user. All these properties are defined within the VEGA system.

---

[3] ViW - Virtuelle Werkstatt (Virtual Workshop Software). This software convert the CAD models to small voxels and exports these data as a polygon model

For the post-modelling, e.g. creation of the virtual workshop or the components that are not available in the CATIA system , the MultiGen [7] software was used.

Figure 9.2 shows the overall process of conversion of the data.

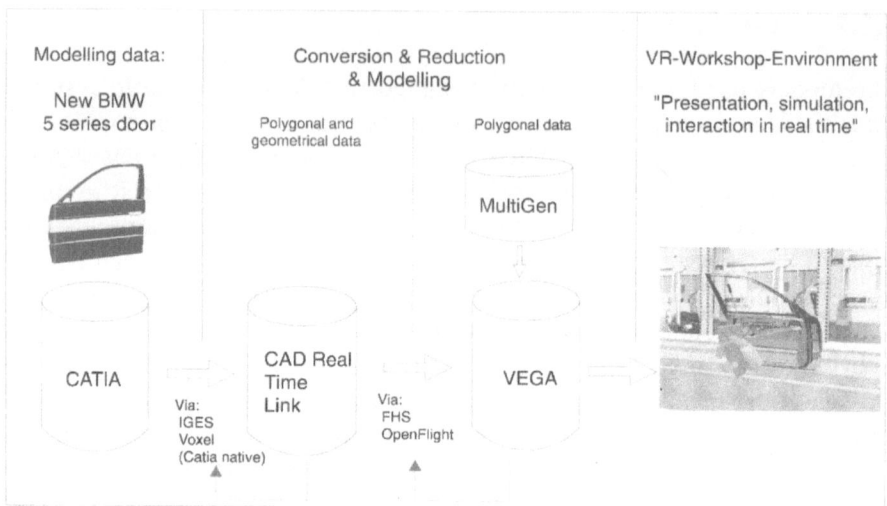

Figure 9.2. Data processing for VEGA VR system

The demonstrator runs on a dual-pipe Onyx RE2 with 12 CPU's, two Multi Channel Options, eight Raster Managers and 768 Mbytes RAM. Two data gloves were used (5th glove, Fifth Dimension Technologies) as input devices. For the gloves it were various gestures were defined for the interaction and navigation in the virtual workshop (e.g. fist - with this gesture it was possible to grasp an object, as soon as the virtual hand collided or touched it). A Head-Mounted Display (VR4 Virtual Research) is used for immersive and stereoscopic views. A big-screen rear stereo projection was used for presentation to the audience.

9.1.3 Procedure

A complete VR workshop with all the characteristics of the real workshop was created. One of these characteristics allows the user to pick-up a special identification card (the ID is on the table outside the workshop) and insert it into the control box on the right side of the door (see Figure 9.3).

Once the check-in has been effected, the door will open and the user can then enter the VR workshop.

Inside, there is a table where the parts (window regulator and door lock) and tools lie and the measuring plate where all investigations, checks and assembly simulations will be processed, (see Figure 9.4).

In the background some real sounds provide the impression of being in an actual workshop. If a collision occurs during installation of the components, a real sound feedback will be produced by the system.

The following functionalities were created to fulfil the assembly and disassembly tasks.

- Picture of workshop environment and tools
- Real-time collision identification (audio and visual)
- Documentation of installation methods as reproducible trace (e.g. record assembly path)
- Creation of the envelope volume, that is to say the space needed to install a component
- Using a screwdriver
- 3D menu (with switches for cameras, tool selection, recording installation travel etc.)
- Simultaneous stereo output on monitor, screen and data helmet (HMD).

Below are some pictures to illustrate the check-in process for the workshop, inside the workshop and the user in the workshop.

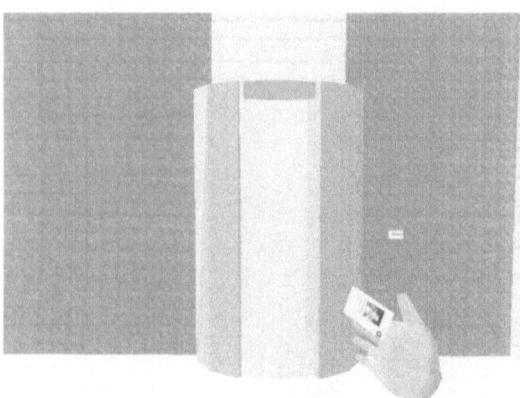

Figure 9.3. Outside the workshop. When the ID is inserted into the control box, it is possible to enter the workshop.

Figure 9.4. Inside the workshop. Parts and tools for assembly simulation.

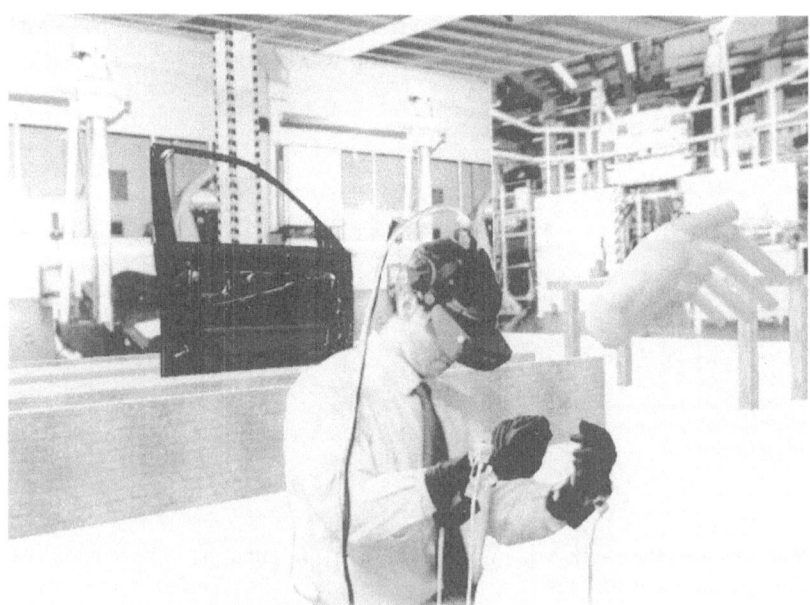

Figure 9.5. Assembly simulation in the virtual workshop environment
(see section Color Plates).

9.2 Project results

The study showed that for people dealing with three-dimensional computer models, VR represents a new experience dimension. Many things which are considered as quite natural in reality have to be trained in virtual worlds, for example movement of components or complex estimation of distances. VR presents the user with a sensation of zero gravity, since gravity influences cannot be simulated (so far).

Technical fringe conditions must of course also be clarified if VR is to be employed as a tool for assembly simulation. So far the users have no realistic perception of the objects that they touch, that is to say, no sensation of the components' surface properties (haptic feedback), nor do they sense any resistance when grasping the components (force feedback). Collision detection in the VR system tested was purely acoustic and visual.

Assembly simulation is only possible if the user receives a realistic feedback of component collisions.

Today's graphic computers, such as the ONXY used here, are not able to render a scene with more than 30-40 thousand polygons in real time (that means 30 frames/sec.). That means that for a complex scenario it is necessary to reduce the precision of the models and use texture-mapping and other algorithms like LOD (Level-of-Detail) to archive the needed real time simulation.

9.3 Future tasks

Today's VR has the potential for initial use in industrial fields. If we succeed in coorperating with research institutes and IT suppliers in bringing this new technology into line with future master build requirements, VR could, if introduced more widely, have a considerable influence on minimising development expenses, for instance by reducing the number of physical mock-ups and auxiliary materials.

To achieve this target, the following functional conditions must be realised:

- Simulation of flexible components (e.g. electrical cables)
- Distance analysis
- Kinematics analysis
- Simulation of components' physical features (e. g. gravity)
- Force feedback

- Haptic feedback

Finally, it is important that ergonomics and usability of all the VR input and output devices should be improved (e.g. HMD without cables, infra-red tracking systems, etc.).

When all these requirements are fulfilled, acceptance and the decision to integrate VR technologies into today's practical work will be guaranteed.

Acknowledgements

We would like to express our thanks to Mr. Peter Reindl (DMU Project Leader at BMW), Mr. Hasmuhk Jina (Rover) and Mr. Stefan Wallner (graduate study trainee at BMW) for their effective and willing cooperation. In this respect we especially appreciate the commitment of the Clarus company in Sweden (Mr. Stefan Hallin and Mr. Jan Grund-Pedersen) in realising the project requirements quickly and making the input and output devices available on loan at short notice. We also would like to thank Mr. Jost Bernasch and his staff (BMW driving simulator) for supplying the necessary computer equipment.

References

[1] AIT DMU project partners, Functional Specification for Digital Mock-up, Project deliverable, June 1995. (restricted document access to AIT Consortium)
[2] AIT DMU project partners, Digital Mock-up Cluster Coordination and Management, Project deliverable, December 1995. (restricted document access to AIT Consortium
[3] CAD Real-Time Link^{TM}, User's Guide, Version 1.1, Prosolvia Clarus AB, Sweden, Nov. 1995.
[4] VEGA, Lynx User's Guide, Version 1.3, Paradigm Simulation Inc., Dallas, USA, 1994.
[5] VEGA, VR User's Guide - VEGA VR Peripheral Module, Version 1.0, Prosolvia Clarus AB, Sweden, 1995.
[6] ViW, Virtuelle Werkstatt standalone-Benutzerhandbuch, Version 2.1, Tecoplan Informatik GmbH, 1995.
[7] MultiGen, User's Guide Version 1.3, Paradigm Simulation Inc., Dallas, USA, 1994
[8] Kress, H.; Schroder, K., Präsentation und Visualisierung für Digital Mock-up; Studie für das AIT Projekt DMU im Auftrag von BMW AG München, 1995.

Color Plates

Figure 1.5. Using Boom for interior design, (see page 21).

Figure 1.6. Openning the hood of a virtual car (AIT, IGD), (see page 23).

Figure 1.8. Assembly simulation with collision response (see page 24).

Figure 1.9. Interactive immersive visualization of a flow field (VW, IGD), (see page 24).

Figure 2.2. Ergonomic study in a maintence simulation (see page 44).

Figure 2.3. Virtaul design model of a Mercedes car with reflection simulation (see page 45).

Figure 3.1. The virtual plane (ViP) from ZGDV (see page 62).

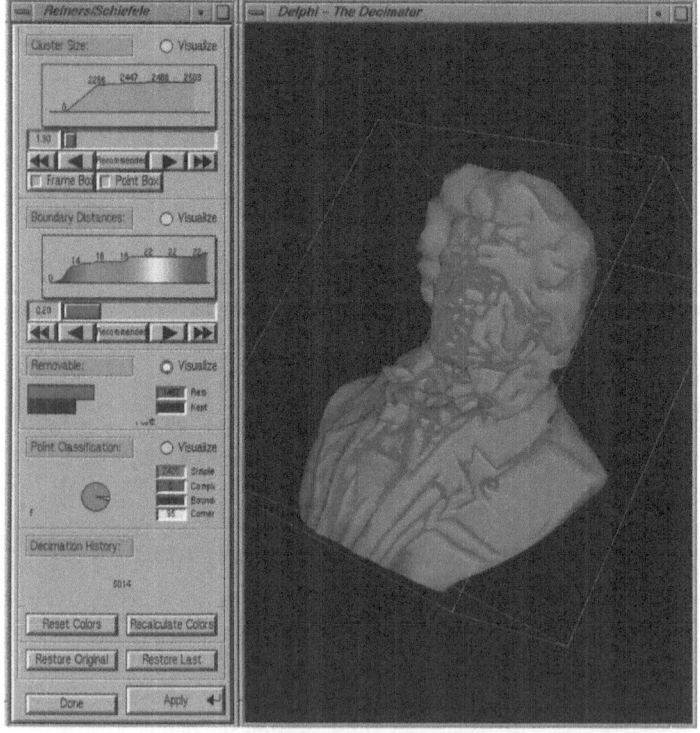

Figure 4.8. Different visualization methods in Delphi (see page 89).

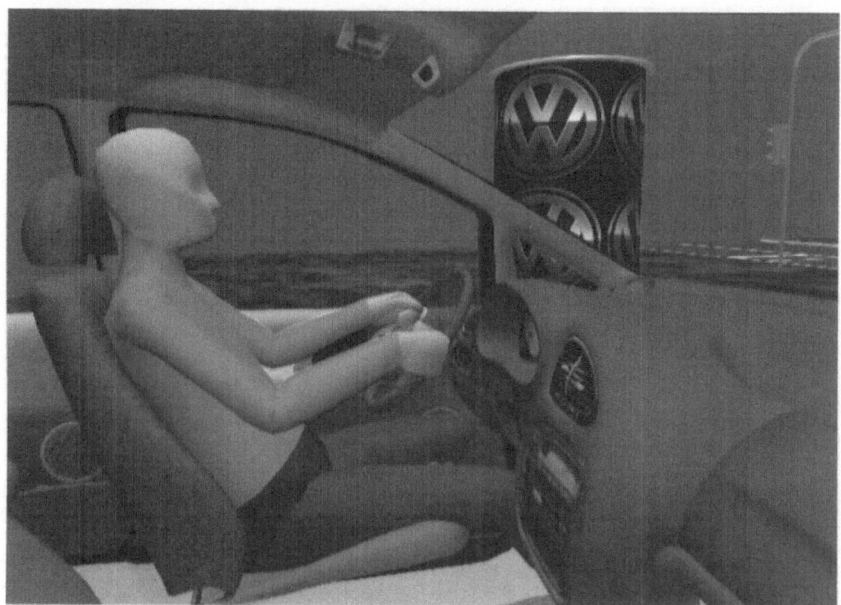

Figure 6.5. Ergnomic analysis (VW), (see page 113).

Figure 6.7. Virtual design model of *Sharan* from Volkswagen (see page 117).

Figure 7.7. Virtual factory simulation (Deneb), (see page 129).

Figure 7.11. A fire fighting training scenario (Deneb), (see page 136).

Figure 8.2. VITAL Virtual Laboratory, (see page 142).

Figure 8.6. WARO virtual scenario, (see page 149).

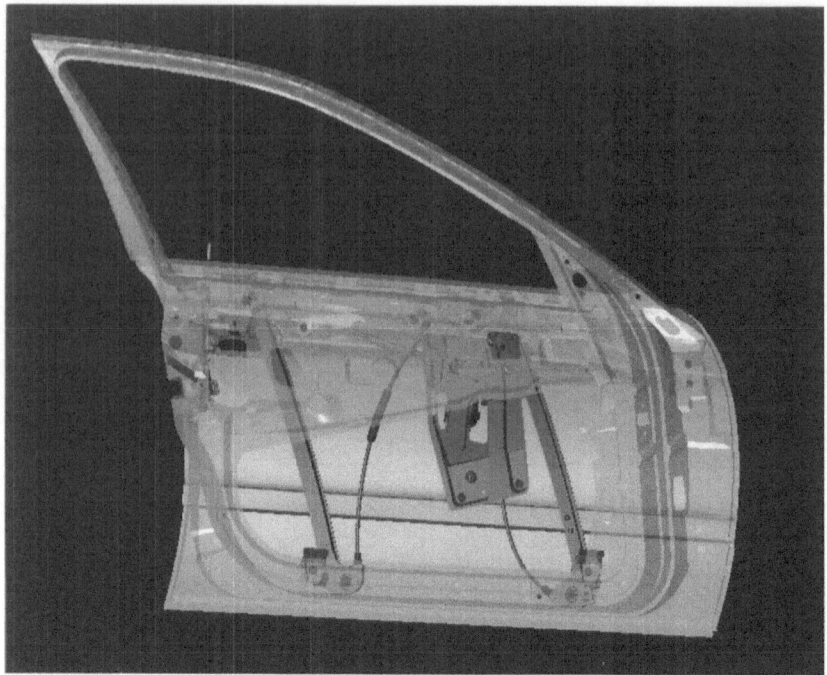

Figure 9.1. Door with intalled lock and window regulator, (see page 152).

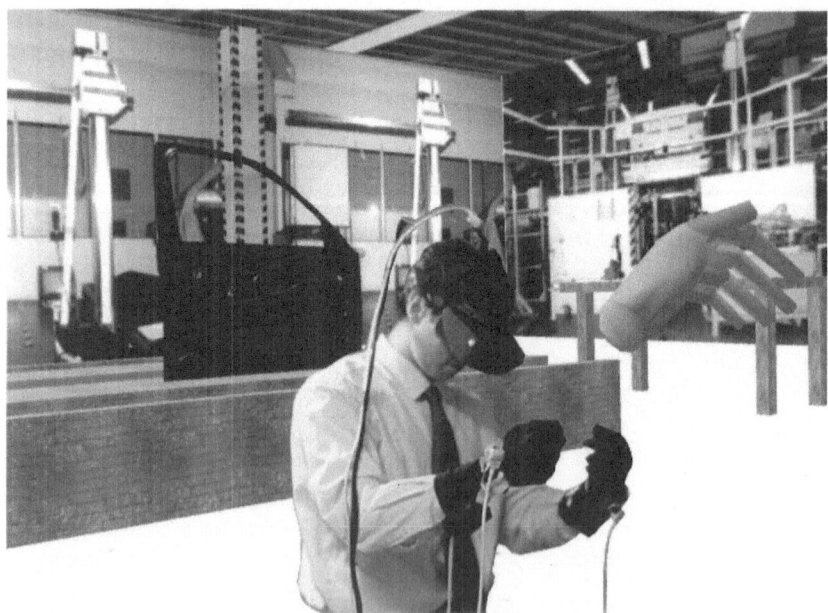

Figure 9.5. Assembly simulation in the virtual workshop enviroment, (see page 156).

List of Figures

Springer
and the
environment